Advances in Intelligent Robotics and Collaborative Automation

RIVER PUBLISHERS SERIES IN AUTOMATION, CONTROL AND ROBOTICS

Volume 1

Series Editors

Tarek Sobh
University of Bridgeport,
USA

André Veltman
PIAK and TU Eindhoven,
The Netherlands

The "River Publishers Series in Automation, Control and Robotics" is a series of comprehensive academic and professional books which focus on the theory and applications of automation, control and robotics. The series focuses on topics ranging from the theory and use of control systems, automation engineering, robotics and intelligent machines.

Books published in the series include research monographs, edited volumes, handbooks and textbooks. The books provide professionals, researchers, educators, and advanced students in the field with an invaluable insight into the latest research and developments.

Topics covered in the series include, but are by no means restricted to the following:

- Robots and Intelligent Machines
- Robotics
- Control Systems
- Control Theory
- Automation Engineering

For a list of other books in this series, visit www.riverpublishers.com
http://www.riverpublishers.com/series.php?msg=Automation,_Control_and_Robotics

Advances in Intelligent Robotics and Collaborative Automation

Editors

Richard Duro

Yuriy Kondratenko

LONDON AND NEW YORK

Published 2015 by River Publishers
River Publishers
Alsbjergvej 10, 9260 Gistrup, Denmark
www.riverpublishers.com

Distributed exclusively by Routledge
4 Park Square, Milton Park, Abingdon, Oxon OX14 4RN
605 Third Avenue, New York, NY 10017, USA

Routledge is an imprint of the Taylor & Francis Group, an informa business

First issued in paperback 2023

Publisher's Note
The publisher has gone to great lengths to ensure the quality of this reprint but points out that some imperfections in the original copies may be apparent.

Advances in Intelligent Robotics and Collaborative Automation / by Richard Duro, Yuriy Kondratenko.

The use of general descriptive names, registered names, trademarks, service marks, etc. in this publication does not imply, even in the absence of a specific statement, that such names are exempt from the relevant protective laws and regulations and therefore free for general use.

The publisher, the authors and the editors are safe to assume that the advice and information in this book are believed to be true and accurate at the date of publication. Neither the publisher nor the authors or the editors give a warranty, express or implied, with respect to the material contained herein or for any errors or omissions that may have been made.

While every effort is made to provide dependable information, the publisher, authors, and editors cannot be held responsible for any errors or omissions.

ISBN 13: 978-87-7022-975-3 (pbk)
ISBN 13: 978-87-93237-03-2 (hbk)
ISBN 13: 978-1-003-33711-9 (ebk)

Contents

Preface

This book provides an overview of a series of advanced trends in robotics as well as of design and development methodologies for intelligent robots and their intelligent components.

All the contributionswere discussed at the International IEEE Conference IDAACS-2013 (Berlin, Germany, 12–14 June, 2013). The IDAACS workshop series is established as a forum for high quality reports on state-of-the-art theory, technology and applications of intelligent data acquisition and advanced computer systems. All of these techniques and applications have experienced a rapid expansion in the last few years that has resulted in more intelligent, sensitive, and accurate methods for the acquisition of data and its processing applied to manufacturing process control and inspection, environmental and medical monitoring and diagnostics. One of the most interesting paradigms that encompass much of the research presented at IDAACs is Intelligent Robotic Systems, and this is the area this book concentrates on.

The success of IDAACS arises not only from the importance of the topics it focuses on, but also because of its nature as a unique forum for establishing scientific contacts between research teams and scientists from different countries. This purpose has become one of the main reasons for the rapid success of IDAACS, as it turns out to be one of the few events in this area of research where Western and former Eastern European scientists can discuss and exchange ideas and information, allowing them to characterize common and articulated research activities and creating the environment for establishing joint research collaborations. It provides an opportunity for all the participants to discuss topics with colleagues from different spheres such as academia, industry, and public and private research institutions. Even though this book concentrates on providing insights into what is being done in the area of robotics and intelligent systems, the papers that were selected reflect the variety of research presented during the workshop as well as the very diverse fields that may benefit from these techniques.

In terms of structure, the 13 chapters of the book are grouped into four sections: Robots, Control and Intelligence, Sensing and Collaborative

Automation. The chapters have been thought out to provide an easy to follow introduction to the topics that are addressed, including the most relevant references, so that anyone interested in them can start their introduction to the topic through these references. At the same time, all of them correspond to different aspects of work in progress being carried out in various laboratories throughout the world and, therefore, provide information on the state of the art of some of these topics.

The first part, "Robots", includes three contributions:

"A Modular Architecture for Developing Robots for Industrial Applications", by A. Faíña, F. Orjales, D. Souto, F. Bellas and R. J. Duro, considers ways to make feasible the use of robots in many sectors characterized by dynamic and unstructured environments. The authors propose a new approach, based on modular robotics, to allow the fast deployment of robots to solve specific tasks. In this approach, the authors start by defining the industrial settings the architecture is aimed at and then extract the main features that would be required from a modular robotic architecture to operate successfully in this context. Finally, a particular heterogeneous modular robotic architecture is designed from these requirements and a laboratory implementation of it is built in order to test its capabilities and show its versatility using a set of different configurations including manipulators, climbers and walkers.

S. Osadchy, V. Zozulya and A. Timoshenko, in "The Dynamic Characteristics of a Manipulator with Parallel Kinematic Structure Based on Experimental Data", studies two identification techniques which the authors found most useful in examining the dynamic characteristics of a manipulator with a parallel kinematic structure as an object of control. These techniques emphasize a frequency domain approach. If all input/output signals of an object can be measured then the first one of such techniques may be used for identification. In the case when all disturbances cannot be measured the second identification technique may be used.

In "An Autonomous Scale Ship Model for Parametric Rolling Towing Tank Testing", M. Míguez González, A. Deibe, F. Orjales, B. Priego and F. López Peña analyze a special kind of robotic system model, in particular, a self-propelled scale ship model for model testing, with the main characteristic of not having any material link to a towing device to carry out the tests. This model has been fully instrumented in order to acquire all the significant raw data, process them onboard and communicate with an inshore station.

The second part "Control and Intelligence" includes four contributions: In "Autonomous Knowledge Discovery Based on Artificial Curiosity Driven Learning by Interaction", K. Madani, D. M. Ramik and C. Sabourin investigate the development of a real-time intelligent system allowing a robot to discover its surrounding world and to learn autonomously new knowledge about it by semantically interacting with humans. The learning is performed by observation and by interaction with a human. The authors provide experimental results both using simulated environments and implementing the approach on a humanoid robot in a real-world environment including everyday objects. The proposed approach allows a humanoid robot to learn without negative input and from a small number of samples.

F. Kulakov and S. Chernakova, in "Information Technology for Interactive Robot Task Training through Demonstration of Movement", consider the problem of remote robot control, which includes the solution of the following routine problems: surveillance of the remote working area, remote operation of the robot situated in the remote working area, as well as pre-training of the robot. Authors propose a new technique for robot control using intelligent multimodal human-machine interfaces (HMI). The application of the new training technology is very promising for space robots as well as for modern assembly plants, including the use of micro-and nanorobots.

In "A Multi-Agent Reinforcement Learning Approach for the Efficient Control of Mobile Robots", U. Dziomin, A. Kabysh, R. Stetter and V. Golovko present a multi-agent control architecture for the efficient control of a multi-wheeled mobile platform. The proposed control architecture is based on the decomposition of a platform into a holonic, homogenous, multi-agent system. The multi-agent system incorporates multiple Q-learning agents, which permits them to effectively control every wheel relative to other wheels. The learning process consists module positioning—where the agents learn to minimize the error of orientation, and cooperative movement—where the agents learn to adjust the desired velocity in order to conform to the desired position in formation. Experiments with a simulation model and the real robot are discussed in details.

D. Oskin, A. Dyda, S. Longhi and A. Monteriù, in "Underwater Robot Intelligent Control Based on Multilayer Neural Network", analyse the design of an intelligent neural network based control system for underwater robots. A new algorithm for intelligent controller learning is derived using the speed gradient method. The proposed systems provide robot dynamics close to the reference ones. Simulation results of neural network control systems for underwater robot dynamics with parameter and partial structural uncertainty have confirmed the perspectives and effectiveness of the developed approach.

The third part "Sensing" includes four contributions:

"Advanced Trends in Design of Slip Displacement Sensors for Intelligent Robots", by Y. Kondratenko and V. Kondratenko, discusses advanced trends in the design of modern tactile sensors andsensor systems for intelligent robots. The detection of slip displacement signals provides information on three approaches for using slip displacement signals, in particular, for the correction of the clamping force, the identification of manipulated object mass and the correction of the robot control algorithm.The chapter presents the analysis of different methods for the detectionof slip displacement signals, as well as new sensor schemes, mathematical models and correction methods.

T. Happek, U. Lang, T. Bockmeier, D. Neubauer and A. Kuznietsov, in "Distributed Data Acquisition and Control Systems for a Sized Autonomous Vehicle", present an autonomous car with distributed data processing. The car is controlled by a multitude of independent sensors. For lane detection, a camera is used, which detects the lane marks using a Hough transformation. Once the camera detects these, one of them is selected to be followed by the car. This lane is verified by the other sensors of the car. These sensors check the route for obstructions or allow the car to scan a parking space and to park on the roadside if the gap is large enough.

In "Polymetric Sensing in Intelligent Systems", Yu. Zhukov, B. Gordeev, A. Zivenko and A. Nakonechniy examine the up-to-date relationship between the theory of polymetric measurements and the state of the art in intelligent system sensing. The chapter discusses concepts of polymetric measurements, corresponding to monitoring information systems used in different technologies and some prospects for polymetric sensing in intelligent systems and robots. The application of the described concepts in technological processes ready to be controlled by intelligent systems is illustrated.

D. Popescu, G. Stamatescu, A. Maciuca and M. Strutu, in "Design and Implementation of Wireless SensorNetwork Based on Multilevel Femtocells for Home Monitoring", propose an intelligent femtocell-based sensor network for home monitoring of elderly or people with chronic diseases. The femtocell is defined as a small sensor network which is placed into the patient's house and consists of both mobile and fixed sensors disposed on three layers. The first layer contains body sensors attached to the patient that monitor different health parameters, patient location, position and possible falls. The second layer is dedicated for ambient sensors and routing inside the cell. The third layer contains emergency ambient sensors that cover burglary events or toxic gas concentration, distributed by necessities. Cell implementation is based on The IRIS family of motes running the embedded software for resource constrained devices, TinyOS. Experimental results within the system architecture are presented for a detailed analysis and validation.

The fourth part "Collaborative automation" includes two contributions:

In "Common Framework Model for Multi-purpose Underwater Data Collection Devices Deployed with Remotely Operated Vehicles", M. Caraivan, V. Dache and V. Sgarciu presents a common framework model for multi-purpose underwater sensors used for offshore exploration. The development of real-time applications for marine operations focusing on modern modeling and simulation methods are discussed with addressing deployment challenges of underwater sensor networks "Safe-Nets" by using Remotely Operated Vehicles.

Finally, S. Gansemer, J. Sell, U. Grossmann, E. Eren, B. Horster, T. Horster-Möller and C. Rusch, in "M2M in Agriculture - Business Models and Security Issues" consider the machine-to-machine communication (M2M) as one of the major ICT innovations. A concept for process optimization in agricultural business using M2M technologies is presented using three application scenarios. Within that concept standardization and communication as well as security aspects are discussed.

The papers selected for this book are extended and improved versions of those presented at the workshop and as such are significantly expanded with respect to the original ones presented at the workshop. Obviously, this set of papers are just a sample of the dozens of presentations and results that were seen at IDAACS 2011, but we do believe that they provide an overview of some of the problems in the area of robotics systems and intelligent automation and the approaches and techniques that relevant research groups within this area are employing to try to solve them. We would like to express our appreciation to all authors for their contributions as well as to reviewers for their timely and interesting comments and suggestions. We certainly look forward to working with you again.

Yuriy Kondratenko
Richard Duro

List of Figures

List of Tables

List of Abbreviations

ABS	Acrylonitrile butadiene styrene
ABS	American Bureau of Shipping
ADC	analog to digital converter
AES	Advanced Encryption Standard
AI	Artificial Intelligence
ANN	Artificial Neural Network
AquaRET	Aquatic Renewable Energy Technologies
ASN	Ambient sensor network
AUV	Autonomous Underwater Vehicle
AWS	Archimedes Waveswing
BCU	Behavior Control Unit
BOP	Blow-Out Preventer
BSN	Body sensor network
CAN	Controller Area Network
CCD	charge-coupled device
CCF	Conscious Cognitive Functions
CCMP	Counter Mode/CBC-MAC Protocol
CCSDBS	computer-aided floating dock ballasting process control and monitoring system
CMF	continuous max-flow algorithm
CO	Carbon monoxide
CO_2	Carbon dioxide
COIL	Columbia Object Image Library
CP	characteristic points
CRL	Certficate Revocation List
CS(SC)	system of coordinates
CSMA/CA	Carrier sense multiple access/collision avoidance
CU	Communication Unit
DB	database
DC	Direct Current
DHCP	Dynamic Host Configuration Protocol
DHSS	Direct hopping spread spectrum
DMA	decision-making agency
DMP	decision-making person

DNS	Domain Name System
ECG	Electrocardiogram
EE	external environment
ESC	Electronic Speed Control
ESN	Emergency sensor network
EU	European Union
FHSS	Frequency hopping spread spectrum
FIM	Fisher Information Matrix
FMN	Femtocell management node
FPSO	Floating production storage and offloading
FSM	frame-structured model
GIS	geographic information systems
GM	graphical model
GM	Ship Metacentric Height
GND	Ground
GPS	Global Positioning System
GSM	Global System for Mobile Communications
GUI	Graphical User Interface
H^2MAS	holonic homogenous multi-agent system
HLAU	High-level Knowledge Acquisition Unit
HMI	human-machine interfaces
HTPS	hand position tracking system
HTTPS	Secure Hypertext Transfer Protocol
ICT	Information and Communication Technology
IE	internal environment
IEEE	Institute of electrical and electronics engineers
IMI	intelligent multimodal interface
IMM	intellectual multy modal system
IMU	Inertial Measurement Unit
INE	information environment agency
IP	Internet Protocol
IRR	ideal rational robot
IS	International System (Measuring)
ISM	Industrial Scientific Medical
ISO	International Standardization Organization
ITTC	International Towing Tank Conference
JRE	Java Runtime Environment
KB	knowledge base
LASCOS	loading and safety control system
LHP	left half-plane
LKAU	Low-level Knowledge Acquisition Unit
LNG	liquefied natural gas

LPG	liquefied petroleum gas
M2M	Machine To Machine
MAS	multi-agent system
MEE	combination of the EE
MEMS	Microelectromechanical Systems
MFM	motion shape models
MIE	particular IE
MIT	Massachusetts Institute of Technology
MiWi	Wireless protocol designed by Microchip Technology
MM	mathematical model
MMI	man-machine interface
MOE	model of the OE
MP	Multilayer Perceptron
MPNN	Multilayer Perceptron Neural Network
NEM	navigation environment monitoring
NN	Neural Network
NU	Navigation Unit
OCSP	Online Certificate Status Protocol
ODA	Operations Data Aquisition
OE	objects of the environment
OpenCV	Open Computer Vision
OPI	operator interface agency
OWC	OscillatingWater Column
PC	Personal computer
PCB	printed circuit board
PCBs	Printed Circuit Boards
PID	Proportional-Integral-Derivative
PPA	polymetric perceptive agency
PT	Process Transparency
RADIUS	Remote Authentication Dias-in User Service
RC	Radio Control
RCS	robot control system
RF	Radio Frequency
RHP	right half-plane
RMS	Root Mean Square
ROV	Remotely (Underwater) Operated Vehicle
RSN	Robust Secure Network
RSU	Remote Software Update
SADCO	systems for automated distant (remote) control
SDS	slip displacement sensor
SIFT	Scale-invariant feature transform
SIM	Subscriber Identity Module

SMA	sensory monitoring agency
SNR	Signal to Noise Ratio
SPM	ship state parameters monitoring
SSM	sea state monitoring
Tbps	Terrabytes per second
TDR	time domain reflectometry
THM	tracking the head motion
TMS	Tether Management System
TSHP	system for tracking hand movements
TSP	telesensor programming
UART	Universal Asynchronous Receiver Transmitter
UAV	Unmanned Aerial Vehicle
UCF	Unconscious Cognitive Functions
UMTS	Universal Mobile Telecommunications System
UPEC	University Paris-Est Creteil
UR	Underwater Robot
USART	Universal Synchronous/Asynchronous Receiver Transmitter
USB	Universal Serial Bus
USD	United States Dollars
UUV	Unmanned Underwater Vehicle
VB.NET	Visual Basic .NET
VSM	virtual ship model
WCM	weather conditions model
WCS	Windows Color System
WIFI	Wireless Fidelity
WLAN	Wireless Local Area Network

1

A Modular Architecture for Developing Robots for Industrial Applications

**A. Faíña[1], F. Orjales[2], D. Souto[2], F. Bellas[2]
and R. J. Duro[2]**

[1]IT University of Copenhagen, Denmark
[2]Integrated Group for Engineering Research, University of Coruna, Spain
Corresponding author: A. Faíña <anfv@itu.dk>

Abstract

This chapter is concerned with proposing ways to make feasible the use of robots in many sectors characterized by dynamic and unstructured environments. In particular, we are interested in addressing the problem through a new approach, based on modular robotics, to allow the fast deployment of robots to solve specific tasks. A series of authors have previously proposed modular architectures, albeit mostly in laboratory settings. For this reason, their designs were usually more focused on what could be built instead of what was necessary for industrial operations. The approach presented here addresses the problem the other way around. In this line, we start by defining the industrial settings the architecture is aimed at and then extract the main features that would be required from a modular robotic architecture to operate successfully in this context. Finally, a particular heterogeneous modular robotic architecture is designed from these requirements and a laboratory implementation of it is built in order to test its capabilities and show its versatility using a set of different configurations including manipulators, climbers and walkers.

Keywords: Modular robots, industrial automation, multi-robot systems.

Advances in Intelligent Robotics and Collaborative Automation, 1–26.

1.1 Introduction

There are several industrial sectors, such as shipyards or construction, where the use of robots is still very low. These sectors are characterized by presenting dynamic and unstructured work environments where the work is not carried out in a chain production line, but rather, the workers have to move to the structures that are being built and these structures change during the construction process. Basically, these are the main reasons to explain the low level of automation in these sectors. Despite this, there are some cases in which robot systems have been considered in order to increase automation in these areas. However, they were developed for very specific tasks, that is, as specialists. Some examples are robots for operations such as grit-blasting [1, 2], welding [3], painting [4, 5], installation of structures [6, 7] or inspection [8, 9]. Nevertheless, their global impact on the sector is still low [10]. The main reason for this low penetration is the high cost of the development of a robot for a specialized task and the large number of different types of tasks that must be carried out in these industries. In other words, it is not practical to have a large group of expensive robots, each one of which will only be used for a particular task and will be doing nothing the rest of the time.

In the last few years, in order to increase the level of automation in the aforementioned environments, several approaches have been proposed based on multi-component robotics systems as an alternative to the use of one robot for each task [11–13]. These approaches seek to obtain simple robotic systems capable of adapting, easily and quickly, to different environments and tasks according to the requirements of the situation.

Multi-component robotic systems can be classified into three categories: distributed, linked and modular robots [14]; however, in this work, only the last category will be taken into account. Thus, we explore an approach based on modular robotics, which basically seeks the re-utilization of pre-designed robotic modules. We want to develop an architecture that with a small set of modules can lead to many different types of robots for performing different tasks.

In the last two decades, several proposals of modular architectures for autonomous robots have been made [15, 16]. An early approach to modular architectures resulted in what was called 'modular mobile robotic systems.' These robots can move around the environment, and they can connect to one another to form complex structures for performing tasks that cannot be carried out by a single unit. Examples are CEBOT [17] or SMC-Rover [18]. Another type of modular architecture is lattice robots. These robots can form compact three-dimensional structures or lattices over which one module or a set of them can move. Atron [19] and Tetrobot systems [20] are examples of this architecture.

A different approach to modularity is provided by the so-called chain-based architecture, examples of which are modular robots such as Polybot [21], M-TRAN [22] or Superbot [23]. This kind of architecture has shown its versatility in several tasks such as carrying or handling payloads, climbing staircases or ropes or locomotion in long tests or in sandy terrains [24–26]. In addition, some of them were designed specifically for dynamic and unstructured environments. This is the case of the Superbot system, which was developed for unsupervised operation in real environments, resisting abrasion and physical impacts, and including enhanced sensing and communications capabilities.

However, and despite the emphasis on real environments, they are mostly laboratory concept testing approaches with an emphasis on autonomous robots and self-reconfigurable systems rather than on. That is, these architectures were not designed to work in industrial settings and, consequently, their components and characteristics were not derived from an analysis of the needs and particularities of these environments. In fact, they are mostly based on the use of a single type of module to simplify their implementation. Additionally, these homogeneous architectures lead to the need of using a large number of modules to perform some very simple tasks.

On the other hand, we can find another expression of modular robotics, which appears as a result of the addition of modularity to robot architectures. An example is modular manipulators which have mostly been studied for their use in industrial environments. These types of manipulators can be re-coupled to achieve, for example, larger load capacities or to extend their workspace. Most of them can obtain a representation of their morphology or configuration and automatically obtain their direct and inverse kinematics and dynamics. There are homogeneous architectures and there are also architectures with modules specialized in different movements but mainly with rotational joints. Nevertheless, they are usually aimed at static tasks [27, 28] and are much less versatile than real complete modular architectures. In this line, companies such as Schunk Intec Inc or Robotics Design Inc. are commercializing products inspired by this last approach. Both companies have developed modular robotic manipulators with mechanical connections, but these manipulators still need an external control unit configured with the arm's topology.

Currently, new research lines have emerged proposing models that take characteristics of the two areas commented above. For example, some research groups have begun to propose complete versatile heterogeneous modular systems that are designed with industrial applications in mind. An example of this approach is the work of [29] and their heterogeneous architecture. These authors propose a heterogeneous architecture, but in its development, they concentrate on using spherical actuators with 3 degrees of freedom and with a small number of attachment faces in each module. Similarly, other authors have proposed the use of a modular methodology to build robots flexibly and quickly with low costs [30]. This architecture is based on two different rotational modules and several end-effectors such as grippers, suckers and wheels or feet. It has shown its strong potential in a wall-climbing robot application [31]. These approaches are quite interesting, but they still lack some of the features that would be desirable from a real industrially usable heterogeneous modular architecture. For instance, the actuator modules in the first architecture are not independent; they need a power and communications module in order to work. The second system only allows serial chain topologies, which reduces its versatility, or the robot is not able to recognize its own configuration in both architectures.

In this chapter, we are going to address in a top-down manner the main features a modular robotic system or architecture needs to display in order to be adequate for operation in dynamic and unstructured industrial environments. From these features, we will propose a particular architecture and will implement a reduced scale prototype of it. To provide an idea of its appropriateness and versatality, we will finally present some practical applications using the prototype modules.

The rest of the chapter is structured as follows: Section 2 is devoted to the definition of the main characteristics the proposed architecture should have to operate in industrial environments and what design decisions will be taken. Section 3 contains different solutions we have adopted through the presentation of a prototype implementation. Section 4 shows different configurations that the architecture can adopt. Finally, Sections 5 and 6 correspond to the introduction of this architecture in real environments and the main conclusions of the chapter, respectively.

1.2 Main Characteristics for Industrial Operation and Design Decisions

Different aspects need to be kept in mind to decide on a modular robotic architecture for operation in a set of industrial environments. On the one hand, it is necessary to determine the types of environments the architecture is designed for and their principal characteristics, the missions the robots will need to perform in these environments and the implications these have on the motion and actuation capabilities of the robots. Obviously, there are also a series of general characteristics that should be fulfilled when considering industrial operation in general. Consequently, we will first start by identifying here the main features and characteristics a modular architecture should display in order to be able to handle a general dynamic and unstructured industrial environment. This provides the requirements to be met by the architecture so that we can address the problem of providing a set of solutions to comply with these requirements. An initial list of required features would be the following:

- Versatility: The system has to allow to easily build a large number of different configurations in order to adapt to specific tasks;
- Fast deployment: The change of configuration or morphology has to be performed easily and in a short time so that robot operation is not disrupted;
- Fault tolerance: In case of the total failure of a module, the robot has to be able to continue operating minimizing the effects of this loss;
- Robustness: The modules have to be robust to allow working in dirty environments and resisting external forces;
- Reduced cost: The system has to be cheap in terms of manufacturing and operating costs to achieve an economically feasible solution;
- Scalability: The system has to be able to operate with a large number of modules. In fact, limits on the number of modules should be avoided.

To fulfil these requirements, a series of decisions were made. Firstly, the new architecture will be based on a modular chain architecture made up of heterogeneous modules. This type of architecture has been selected because it is well known that it is the general architecture that maximizes versatility. On the other hand, using homogeneous modules is the most common option in modular systems [15, 16, 21–23], because it facilitates module reuse. However, it also limits the range of possible configurations and makes the control of the robot much more complex. In the types of tasks we are considering here, there are several situations that would require a very simple module (e.g., a linear displacement actuator), but which would be very difficult (complex morphology), or even impossible in some cases, to obtain using any of the homogeneous architectures presented. Thus, for the sake of flexibility and versatility, we have chosen to use a set of heterogeneous modules (specialized modules for each type of movement). This solution makes it easier for the resulting robots to perform complex movements as complex kinematic chains can be easily built by joining a small set of different types of modules. Moreover, each module was designed to perform a single basic movement, that is, only one degree of freedom is allowed. This permits using simple mechanisms within the modules, which increases the robustness of the system and reduces the operating and manufacturing costs.

Having decided on the nature of the architecture, now the problem is to decide what modules would be ideal in terms of having the smallest set of modules that covers all possible tasks in a domain. In addition, it should be taken into account that the number of different types of modules needs to be low in order to accomplish the scalability and reduced production cost requirements. To do this, we chose to follow a top-down design strategy. To this end, we studied some typical unstructured industrial environments (shipyards) and defined a set of general missions that needed automation. These missions were then subdivided into tasks and these into operations or sub-tasks that were necessary. From these we deduced the kinematic pairs and finally a simple set of actuator and end-effector modules that would cover the whole domain was obtained. This approach differentiates the architecture presented here from other systems, which are usually designed with a bottom-up strategy (the modules are designed as the first step and then the authors try to figure out how they can be applied).

We have only considered five general types of modules in the architecture:

- Actuators: Modules with motors to generate the robot's motions;
- Effectors: Modules to interact with the environment such as magnets, suckers or grippers;
- Expansion: Modules that increase computational capabilities, memory or autonomy through batteries;
- Sensors: Modules to measure and obtain data from the environment such as cameras and infrared or ultrasonic sensors;
- Linkers: Modules used to join other modules mechanically.

The architecture incorporates these five types of modules, but in this work, we have focused only on the actuator modules. They are the ones around which the morphological aspects of the robots gravitate, and we only employ other modules when strictly necessary to show application examples. Therefore, each module includes a processing unit, one motor, a battery, capabilities to communicate with other modules and the necessary sensors to control its motions. This approach permits achieving a fast deployment of functional robots and their versatility as compared to cases where they require external control units.

The process followed to decide on the different actuator modules corresponds with a top-down design process as presented in Figure 1.1. As a first step, we have considered three basic kinds of general mission the modular robot could accomplish. These are the surface, linear and static missions (top layer). Surface missions are those related with tasks requiring covering any kind of surface (like cleaning a tank). Linear missions are those implying a linear displacement (like weld inspection) and Static missions are those where the robotic unit has a fixed position (like an industrial manipulator).

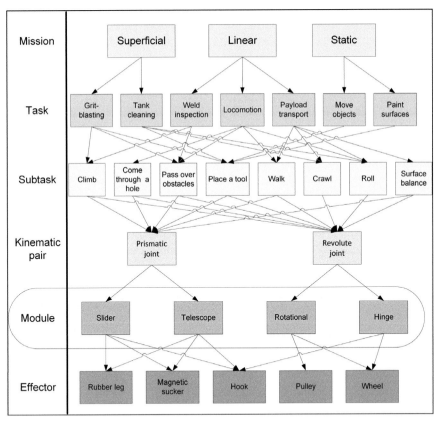

Figure 1.1   Diagram of the selected missions, tasks and sub-tasks considered, and the required actuators and effectors.

The next layer shows the set of possible particular tasks we have considered as necessary according to the previous types of mission, such as grit-blasting, tank cleaning etc. The sub-task layer represents the low-level operations the modular system must carry out to accomplish the task of the previous layer. The next layer represents the kinematic pairs that can be used to perform all the sub-tasks of the last layer. As mentioned above, these pairs only have one degree of freedom. In this case, we have only chosen two kinds of kinematic pairs: prismatic and revolution joints. Nevertheless, each joint was implemented in two different modules in order to specialize the modules to different motion primitives. For the prismatic joint, we have defined a telescopic module with a contraction/expansion motion and a slider module with a linear motion over its structure. The revolution joint also leads to two specialized modules: a rotational module where the rotational axis goes through the two parts of the module, like in wheels or pulleys, and a hinge module. Finally, in the last layer we can see five examples of different effector modules.

Once the actuator modules have been defined, we have to specify the shape or morphology and the connecting faces of each module. Also, and again to increase the versatility of the architecture, each module has been endowed with a large number of attachment faces. This also permits reducing the number of mechanical adapters needed to build different structures. The distribution of the attachment faces will be located on cubic nodes or connection bays within each module. This solution allows creating complex configurations, even closed chains, with modules that are perpendicular, again increasing the versatility of the architecture.

These mechanical connections have to be easily operated in order to allow for the speedy deployment of different configurations. To this end, each attachment face has been provided with mechanisms for transmitting energy and communications between modules in order to avoid external wires. We have also included mechanisms (proprioceptors) that allow the robot to know its morphology or configuration, that is, what module is attached to what face. This last feature is important because it allows the robot to calculate its direct and inverse kinematics and dynamics in order to control its motion in response to high-level commands from an operator.

The robots developed have to be connected to an external power supply with one cable to guarantee the energy needed by all the actuators, effectors and sensors. Nevertheless, the energy is shared among the modules to avoid wires form module to module. In addition, each module contains a small battery to prevent the risk of failure by a sudden loss of energy. These batteries, combined with the energy bus between the modules, allow the robot to place itself in a secure state, maximizing the fault tolerance and the robustness of the system.

Finally, for the sake of robustness, we decided that the communications between modules should allow three different communication paths: a fast and global channel of communications between all the modules that make up a robot, a local channel of communications between two attached modules and a global and wireless communication method. These three redundant channels allow efficient and redundant communications, even between modules that are not physically connected or when a module in the communications path has failed.

Summarizing, the general structure of a heterogeneous modular robotic architecture has been obtained from the set of requirements imposed by operation in an industrial environment and the tasks the robots must perform within it. It turns out that given the complexity of shipyard environments, on which the design was based, the design decisions that were made have led to an architecture that can be quite versatile and adequate for many other tasks and environments. In the following section, we will provide a more in-depth description of the components of the architecture and their characteristics as they were implemented for tests.

1.3 Implementation of a Heterogeneous Modular Architecture Prototype

In the previous section, the main features and components of the developed architecture were presented. Here we are going to provide a description of the different solutions of actuator modules we have adopted through the presentation of a prototype implementation. Throughout this section, the design and morphology of the modules will be explained as well as the different systems needed for it to operate, such as the energy supply system, communications, control system, etc.

Figure 1.2 displays some of the different types of modules that were developed. On the left part, it presents some of the effectors, on the top a linker, a slider on the right and a rotational module, a telescopic module and a hinge in the center. The different modules (actuators, linkers and effectors) have been fully designed and a prototype implementation has been built for each one of them. They all comprise nodes built using fiber glass from milled printed circuit boards (PCBs). These parts are soldered to achieve a solid but lightweight structure. Each module is characterized by having one or more

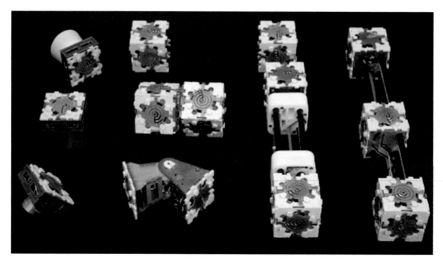

Figure 1.2 Different types of modules developed in this project: three effectors on the left part, a linker on the top, a slider on the right, and in the middle there is a rotational module, a hinge module and a telescopic module.

nodes, which act as connections bays. The shape of the nodes varies depending on the type of module (e.g., it is a cube for the nodes of the slider and telescopic modules). All of the free sides of these nodes provide a connection mechanism that allows connecting them to other modules. The size of the nodes without the connection mechanism is 48x48x48 mm; it is 54x54x54 mm including the connectors.

1.3.1 Actuator Modules

To develop the prototype of the architecture, four different types of actuator modules have been built in accordance to the main features of the architecture described in the previous section. The modules only present one degree of freedom in order to increase robustness and they have different types of joints so that it is easy to build most of the kinematic chains used by real robotic systems in industry. To this end, two linear actuators (slider and telescopic modules) and two rotational actuators (rotational and hinge modules) have been developed. In the case of linear actuators, the slider module has a central node capable of a linear displacement between the end nodes. Any other module can be connected to this central node. The telescopic module only has two nodes and the distance between them can be modified.

On the other hand, the rotational modules have two nodes and allow their relative rotation. These modules are differentiated by the position of the rotation shaft. Whereas the rotational axis of the rotation module goes through the center of both modules, in the hinge it is placed in the union of both nodes and perpendicularly to the line connecting their centers. The main characteristics of the actuator modules are described in Table 1.1.

1.3.1.1 Slider module

This module has two end nodes that are joined together using three carbon fiber tubes and an additional node that slides along the tubes between the end nodes. The distance between the end nodes is 249 mm and the stroke of the slider node is 189 mm. One of the end nodes has a servo with a pulley, which moves a drive belt. The node on the other end has the return pulley and the slider node is fixed to the drive belt. The central node contains the electronics of the module, with power and data wires connecting it to one of the end nodes. There is also a mechanism that coils the wires to adapt them to the position of the slider node.

1.3.1.2 Telescopic module

The telescopic module has two nodes and the distance between them can increase or decrease. Each node has two carbon fiber tubes attached to it. There is an ABS plastic part at the end of the tubes. These parts have two holes with plain bearings to fit the tubes of the other node. One node contains a servo with a drive pulley and the return pulley is in the ABS part of this node. The drive belt that runs in these pulleys is connected to the ABS part of the opposite node. The other node has the electronic board.

1.3.1.3 Rotational module

This module has two nodes that can rotate with respect to each other. A low friction washer between the nodes and a shaft prevents misalignments. One

Table 1.1   Actuator Modules

	Slider	Telescopic	Rotational	Hinge
Type of movement	Linear	Linear	Rotational	Rotational
Stroke	189mm	98mm	360° (1 turn)	200°
N° nodes	3	2	2	2
N° connection faces per node	5–4–5	5–5	5–5	1–1
Weight	360g	345g	250g	140g

node carries a servo with a gear that engages another gear coupled to the shaft. The reduction ratio is 15:46. The servo is modified and its potentiometer is outside attached to a shaft that is operating at a 1:2 ratio with respect to the main shaft. This configuration permits rotations of the module of 360º.

1.3.1.4 Hinge module

This module does not have any connection bay in its structure, only one connection mechanism in each main block. A shaft joins two main parts built from milled PCBs. These parts rotate relative to each other. The reduction of the servo to the shaft is 1:3. The potentiometer of the servo is joined to the shaft to sense the real position of the module.

1.3.2 Connection Mechanism

Different types of physical couplings between modules can be found in the literature, including magnetic couplings, mechanical couplings or even shape memory wires. In this work, we have decided to use a mechanical connection due to the high force requirements in some tasks and due to the power consumption of other options, like in the case of magnetic couplings.

Several mechanical connectors have been developed for modular robots, but most designers focus their efforts in the mechanical aspects, paying less attention to power transmission and communications between modules. Here we have designed a connection mechanism that is able to join two modules mechanically and, at the same time, transmit power and communications. Currently, the connector is manually operated, but its automation is under development.

The connector design can be seen in Figure 1.2 and it has two main parts: a printed circuit board and a resin structure. The resin structure has four pins and four sockets to allow four connections in a multiple of 90 degrees like in [16] and [27]. Inside the resin structure, there is a PCB that can rotate 15 degrees. The PCB has to be forced to fit inside the resin structure, so the PCB remains fixed. When two connectors are faced, the rotation of the PCB of one connector blocks the pins of the other one, and vice versa. The space between the pins of the two connectors is the same as the thickness of the two connector PCBs.

The PCB has four concentric copper tracks on the top side. A mill breaks these tracks in order to provide a cantilever. A small quantity of solder is deposited in the end of the cantilever track. When two connectors are attached, this solder forces the cantilever tracks to bend, so a force is generated. This force maintains the electrical contacts fixed even under vibrations.

Two of the tracks are wider than the other two because they are employed to transmit power (GND and +24V). The other two tracks are employed to transmit data: a CAN bus and local asynchronous communications. The local asynchronous communications track in each connector is directly connected to the microcontroller, while the other tracks are shared for all the connectors of the module. To share these tracks in the node, we choose a surface mount and insulating displacement connector placed at the bottom of the PCB. This solution is used to serially connect the PCBs of the node together in a long string and it allows two modules on the same robot to communicate even in the case of a failure in a module in the path of the message.

1.3.3 Energy

A need for the modular system to require a wire or tether to obtain power or perform communications would limit the resulting robots' motions and their independence. Therefore, one aim of this work is for the architecture to allow for fully autonomous modular robots. This is achieved by means of the installation of batteries in each module and, when the robot needs more power, expansion modules with additional batteries can be attached to it. However, in industrial environments it is often the case that the tools the robots need to use do require cables and hoses to feed them (welding equipment, sandblasting heads, etc.) and, for the sake of simplicity and length of time the robot can operate, it makes a lot of sense to use external power supplies. For this reason, the architecture also allows for tethered operation when this is more convenient, making sure that the power line reaches just one of the modules and then it is internally distributed among the rest of the modules.

The modules developed in this work are powered at 24V, but each module has its own dc converter to reduce the voltage to 5V to power the servomotors and the different electronic systems embedded in each module.

1.3.4 Sensors

All of the modules contain specific sensors to measure the position of their actuator. To this end, the linear modules have a quadrature encoder with 0.32 mm accuracy in their position. The rotational modules are servo controlled, so it is not necessary to know the position of the module. But, in order to improve the precision of the system, we have added a circuit that senses the value of the potentiometer after applying a low pass filter.

Furthermore, all the modules have an accelerometer to provide their spatial orientation. In addition, the local communications established in each attachment face permit identifying the type and the face of the module that is connected to it. This feature, combined with the accelerometer, allows determining the morphology and attitude of the robot without any external help.

All the above-mentioned sensors are particular to each individual module. It means that they only get the data from the module as well. Nevertheless, to perform some tasks (welding, inspection, measuring, etc.), it is necessary to provide to the robot with specific sensors such as camera, ultrasound sensor or whatever. These specific sensor modules are attached to the actuator module that requires it. They are basically nodes (morphologically similar to the rest of the nodes in most modules) with the particular sensor and the processing capabilities to acquire and communicate the data from the particular sensor.

1.3.5 Communications

One of the most difficult tasks in modular robotics is the design of the communications systems (local and global). On the one hand, it has to ensure the adequate coordination between modules, and on the other hand, it has to be able to respond quickly to possible changes in the robot's morphology. That is, it has to adapt when a new module is attached, unattached or even when one module fails. The robot's general morphology has to be detected through the aggregation of the values of the local sensing elements in each module as well as the information they have on the modules they are linked to. For this, we use an asynchronous local communications line for inter-module identification (morphological proprioception).

On the other hand, a CAN bus is used for global communications. It allows performing tasks requiring a critical temporal coordination between remote modules. Also, a MiWi wireless communications system is implemented as a redundant system that is used when we have isolated robotic units or when the CAN bus is saturated.

Additionally, all the modules, except the rotational one, have a micro-USB connection to allow communications to an external computer. This feature and a boot loader allow us to employ a USB memory to load the program without the use of a programmer for microcontrollers. Figure 1.3 shows the printed circuit board (PCB) of the slider module containing all the communications elements.

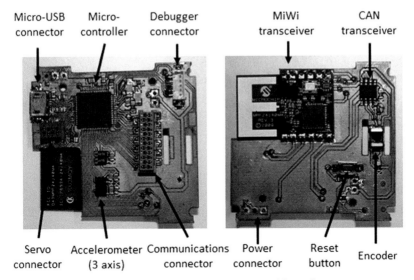

Figure 1.3 Control board for the slider module and its main components.

1.3.6 Control

The system control is responsible for controlling and coordinating all the local tasks within each module, as well as the behaviour of the robot. To do this, in this work, each module carries its own electronics board with its micro-controller (PIC32MX575F512) and a DC/DC converter for power supply. The micro-controller is responsible of the low-level tasks of the module: controlling the actuator, managing the communications stacks and measuring the values of its sensors. As each actuator module has its own characteristics (number of connection faces, encoder type, etc.) and the available space inside the modules is very limited, we have developed a specific PCB for each kind of actuator module. As an example, Figure 1.3 shows the top and bottom side of the control board for the slider module.

Besides the low-level tasks, this solution permits choosing the type of control to be implemented: centralized or distributed. While in a distributed control scheme, each of the modules contributes to the final behaviour through the control of its own actions depending on its sensors or communications to other modules. In a centralized control scheme, one of the modules would be in charge of controlling the actions of all the other modules, with the advantage of having redundant units in case of failure. Additionally, all modules employ the CAN bus to coordinate their actions and to synchronize their clocks. Obviously, this architecture allows for any intermediate type of control scheme.

1.4 Some Configurations for Practical Applications

In this section, we will implement some example configurations using the architecture to show how easy it is to build different types of robots as well as how versatile the architecture is. For the sake of clarity and in order to show the benefits of a heterogeneous architecture, we will show simple configurations developed with only a few modules (videos of these configurations can be found in vimeo.com/afaina/ad-hoc-morphologies).

All the experiments were carried out with the same setup. First, the modules were manually assembled in the configuration to test and we connected one cable for power supply and an USB cable to connect one module to a laptop. After powering up the system, the module that communicates to the laptop is selected as a master module. This master module uses the CAN bus to find other connected modules. Then, it uses the asynchronous communications and the orientation of each module to discover the topology of the robot.

1.4.1 Manipulators

One of the most important pillars of industrial automation are manipulators. Traditional manipulators present a rigid architecture, which complicates their use in different tasks, and they are very heavy and big to be transported in dynamic and unstructured environments. Nevertheless, modular manipulators can be very flexible as they can be entirely reconfigured to adapt to a specific task and the modules can be transported easily across complex environments and then they can be directly assembled on the workplace.

The configuration choice of the manipulator is highly application dependent and it is mostly determined by the workspace shape and size, as well as other factors such as the load to be lifted, the required speed, etc. For instance, the different types of modules in the architecture can also be used to easily implement spherical or polar manipulators. These type of manipulators present a rotational joint at their base and a linear joint for the radial movements as well as another rotational joint to control their height. Thus, a spherical manipulator is constructed using just five modules as shown in the pictures of Figure 1.4. This robot has a magnetic effector to adhere to the metal surface: a rotational module, a hinge module and a prismatic module for motion and a

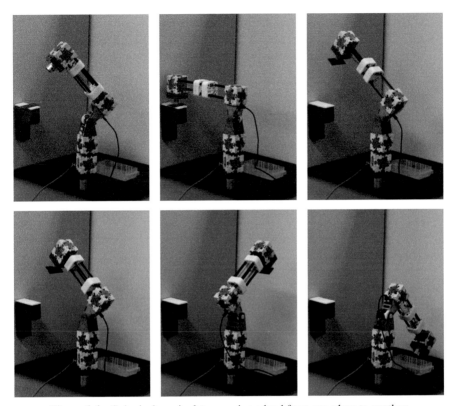

Figure 1.4 Spherical manipulator moving a load from one place to another.

final magnetic effector to manipulate metal pieces. We can see how the robot is able to take an iron part using the electromagnet placed at the end of the manipulator and carry it to another place. The whole process takes around 10 seconds.

Another very common type of manipulator is the cartesian robot. They are constructed using just linear joints and are characterized by a cubic workspace. The ease with which it is possible to produce speed and position control mechanisms for them, their ability to move large loads and their great stability are their major advantages.

An example of a very simple and fully functional cartesian robot is displayed on the left image of Figure 1.5. It is constructed using only two linear modules and a telescopic module for the implementation of its motions, two magnetic effectors to adhere to the metal surface and a smaller magnet that is used as a final effector. The two large magnets used to adhere the robot

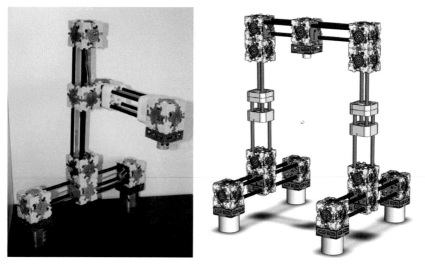

Figure 1.5   Cartesian manipulators for Static missions.

to the metal surface provide better stability than the previous spherical robot and reduce the vibrations on the small magnetic end-effector. In addition, we could implement a gantry style manipulator, as we can observe on the right image of Figure 1.5. This gantry manipulator has great stability as it uses four magnets to adhere to the surface and provides a very stable structure to achieve a high accuracy positioning of its end-effector. Furthermore, this implementation can lift and move heavier loads as it has two pairs of modules working in parallel.

1.4.2 Climber and Walker Robots

The most appropriate configurations to carry small loads or sensors and to move the robots themselves to the workplace are the so-called climber or walker robot configurations. Modular robots should be able to get to hard to reach places and, more importantly, their architecture should allow for their reconfiguration into appropriate morphologies to move through different types of terrains, different sized tunnels or over obstacles. This reconfigurability allows reaching and working in areas where it would be impossible for other kinds of robots to operate. Consequently, being able to obtain simple modular configurations that allow for these walking or climbing operations is important, and in this section we will describe three configurations using our architecture that allow for this.

One of the most prototypical configuration in modular robots is a serial chain of hinge modules. It is called a worm configuration and can be employed for inspection tasks inside pipes using a camera or to pass through narrow passages. Here, we show on Figure 1.6 that we can achieve a working worm type robot using two hinge modules of our architecture. The whole sequence takes around 8 seconds, but the robot's speed could be increased if we use a worm configuration with more modules.

Another example of how using this architecture a functional robot climber can be constructed with just with a few modules is the linear wall climber. This robot consists in a combination of a slider module for motion and two magnet effectors to stick to the metal surface. This simple robot, which is displayed on Figure 1.7. (left), can be used on tasks like measuring ship rib thickness or inspecting a linear weld.

Obviously, the linear climber is unable to avoid obstacles or to turn. Thus, a possibility to achieve configurations with greater capabilities is to use a few more modules. A wall climber robot is shown in Figure 1.7 (right). It can be constructed through the combination of two slider modules, each one of them with two magnetic effectors to adhere to the metal surface, a linear module and

Figure 1.6   A snake Robot that can inspect inside a pipe.

Figure 1.7   Climber and Walker Robots for linear and surface missions.

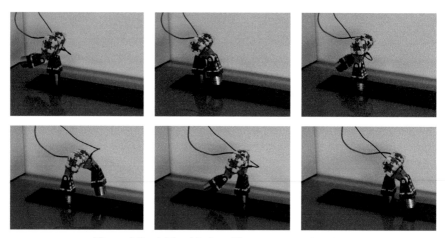

Figure 1.8   A Biped Robot able to overpass obstacles.

a hinge module between them. This configuration allows the robot to move and to turn, making it useful for surface inspection tasks performed with an ultrasonic sensor or other final effectors.

Approximations that are more complex can be created with better locomotion capabilities using other sets of modules. For example, a well-known way to move through an environment is by walking. This way of moving also allows stepping over small obstacles or irregularities. A very simple implementation of a walking robot is shown in Figure 1.8. This configuration is made up of two hinge modules, each one of them with a magnetic effector, joined together by a rotational module. This biped robot is capable of walking over irregular surfaces, stepping over small obstacles and even of moving from a horizontal to a slanted surface.

1.5 Towards Industrial Applications

In this work, we have analyzed the main features and characteristics that a modular architecture should display in order to be able to handle a general dynamic and unstructured industrial environment: versatility, fast deployment, fault tolerance, robustness, reduced cost and scalability. Currently, modular commercial systems have achieved a good fault tolerance, robustness and reduced cost, but they still lack versatility to operate in dynamic industrial environments and their deployment needs at least some hours.

Here, we have developed a new modular architecture taking into account these dynamic environments. An initial analysis has shown that some important features for an architecture of this type is that it should be an heterogeneous modular architecture with a high number of standardized connection faces, different channels of communication, common power buses, and an autonomous and independent control for each module or the robot's ability to discover its morphology. In order to test the architecture in useful tasks, we have implemented some modular prototypes. The results show that we can deploy complex robots for specific tasks in a few minutes and they can be easily controlled through a GUI in a laptop. Furthermore, we can deploy different configurations for similar tasks where we can increase the stability and accuracy of the robot's end-effector using parallel robots.

An industrial implementation of this architecture is still in a development stage, but it will allow working reliably in dynamic and unstructured environments. It will have the same features of our architecture but with an industrial-oriented implementation. The main changes will affect to the robustness of the modules and the connectors. First, modules will be able to support loads and momentums generated by the most typical configurations. In addition, it will be ruggedized to work in real environments, which can present dust or humidity. Regarding the connectors, they will be able to support high loads and, at the same time, they will allow the fast deployment of the robot configurations. We can find one connector with these characteristics in [32] but, additionally, they will have to distribute the energy and communications buses. As the robots have to work in environments with a high presence of ferromagnetic material, such as shipyards, we cannot use the magnetometer values to calculate the relative orientation of the module. Therefore, we will include a sensor to measure the relative orientation between the modules. Finally, one important issue to address is the security of the operators who work near the robots. Most industrial robots are designed to work in close environments with a shield for the worker's protection. Thereby, our modules will have to be compliant for security reasons. This solution is currently used by some companies that sell compliant robots able to work in the presence of humans [33].

1.6 Conclusions

A new heterogeneous modular robotic architecture has been presented which permits building robots in a fast and easy way. The design of the architecture is based on the main features that we consider, in a top-down fashion, that a modular robotic system must have in order to work in industrial environments. A prototype implementation of the architecture was created through the construction of a basic set of modules that allows for the construction of different types of robots. The modules provide for autonomous processing and control, one degree of freedom actuation and a set of communications capabilities so that, through their cooperation, different functional robot structures can be achieved. To demonstrate the versatility of the architecture, a set of robots was built and tested for simple operations, such as manipulation, climbing or walking. Obviously, this prototype implementation is not designed to work in real industrial environments. Nevertheless, the high level of flexibility achieved with very few modules shows that this approach is very promising. We are now addressing the implementation of the architecture in more rugged modules that allow testing in realistic environments.

References

[1] C. Fernandez-Andres, A. Iborra, B. Alvarez, J. Pastor, P. Sanchez, J. Fernandez-Merono and N. Ortega, 'Ship shape in Europe: cooperative robots in the ship repair industry', Robotics & Automation Magazine, IEEE, vol. 12, no. 3, pp. 65–77, 2005.

[2] D. Souto, A. Faiña, A. Deibe, F. Lopez-Peña and R. J. Duro, 'A robot for the unsupervised grit-blasting of ship hulls', International Journal of Advanced Robotic Systems, vol. 9, no. 82, 2012.

[3] G. de Santos, M. Armada and M. Jimenez, 'Ship building with rower', IEEE Robotics & Automation Magazine, vol. 7, no. 4, pp. 35–43, 2000.

[4] B. Naticchia, A. Giretti and A. Carbonari, 'Set up of a robotized system for interior wall painting', in Proceedings of the 23rd International Symposium on Automation and Robotics in Construction (ISARC), pp. 3–5, 2006.

[5] Y. Kim, M. Jung, Y. Cho, J. Lee and U. Jung, 'Conceptual design and feasibility analyses of a robotic system for automated exterior wall painting', International Journal of Advanced Robotic Systems, vol. 4, no. 4, pp. 417–430, 2007.

[6] S. Yu, S. Lee, C. Han, K. Lee and S. Lee, 'Development of the curtain wall installation robot: Performance and efficiency tests at a construction site', Autonomous Robots, vol. 22, no. 3, pp. 281–291, 2007.

[7] P. Gonzalez de Santos, J. Estremera, E. Garcia and M. Armada, 'Power assist devices for installing plaster panels in construction', Automation in Construction, vol. 17, no. 4, pp. 459–466, 2008.

[8] C. Balaguer, A. Gimenez and CM Abderrahim. 'Roma robots for inspection of steel based infrastructures', Industrial Robot: An International Journal, vol. 29, no. 3, pp. 246–251, 2002.

[9] J. Shang, T. Sattar, S. Chen and B. Bridge. 'Design of a climbing robot for inspecting aircraft wings and fuselage', Industrial Robot: An International Journal, vol. 34, no. 6, pp. 495–502, 2007.

[10] T. Yoshida, 'A short history of construction robots research & development in a Japanese company', in Proceedings of the International Symposium on Automation and Robotics in Construction, 2006, pp. 188–193.

[11] C. A. C. Parker, H. Zhang and C. R. Kube, 'Blind bulldozing. Multiple robot nest construction', In IEEE/RSJ International Conference on Intelligent Robots and Systems, (IROS 2003), vol. 2, pp. 2010–2015, 2003.

[12] N. Correll and A. Martinoli, 'Multirobot inspection of industrial machinery', Robotics & Automation Magazine, IEEE, vol. 16, no. 1, pp. 103–112, 2009.

[13] A. Breitenmoser, F. Tâche, G. Caprari, R. Siegwart and R. Moser, 'Magnebike: toward multi climbing robots for power plant inspection', In Proceedings of the 9th International Conference on Autonomous Agents and Multiagent Systems: Industry track, pp. 1713–1720. International Foundation for Autonomous Agents and Multiagent Systems, 2010.

[14] R. J. Duro, M. Graña and J. de Lope, 'On the potential contributions of hybrid intelligent approaches to multicomponent robotic system development', Information Sciences, vol. 180 no. 14, pp. 2635–2648, 2010. Special Section on Hybrid Intelligent Algorithms and Applications.

[15] M. Yim, W. M. Shen, B. Salemi, D. Rus, M. Moll, H. Lipson et al., 'Modular self-reconfigurable robot systems', IEEE Robotics and Automation Magazine, vol. 14, no. 1, pp. 43–52, 2007.

[16] K. Stoy, D. Brandt and D. J. Christensen, 'Self-reconfigurable robots: An Introduction', MIT Press, 2010.

[17] T. Fukuda, S. Nakagawa, Y. Kawauchi and M. Buss, 'Self-organizing robots based on cell structures-cebot', In IEEE International Workshop on Intelligent Robots, pp. 145–150, 1988.

[18] A. Kawakami, A. Torii, K. Motomura and S. Hirose, 'Smc rover: planetary rover with transformable wheels', In SICE 2002. Proceedings of the 41st SICE Annual Conference, vol. 1, pp. 157–162 vol.1, 2002.

[19] M. W. Jorgensen, E. H. Ostergaard and H. H. Lund, 'Modular atron: Modules for a self-reconfigurable robot', In IEEE/RSJ International Conference on Intelligent Robots and Systems, pp. 2068–2073, 2004.

[20] G. J. Hamlin and A. C. Sanderson, 'Tetrobot modular robotics: prototype and experiments'. In Intelligent Robots and Systems '96, IROS 96, Proceedings of the 1996 IEEE/RSJ International Conference on, vol. 2, pp. 390–395, 1996.

[21] M. Yim, D. Duff and K. Roufas, 'Polybot: a modular reconfigurable robot', in IEEE International Conference on Robotics and Automation, vol. 1. IEEE; pp. 514–520, 2000.

[22] S. Murata, E. Yoshida, A. Kamimura, H. Kurokawa, K. Tomita and S. Kokaji, 'M-tran: Self-reconfigurable modular robotic system', IEEE/ASME transactions on mechatronics, vol. 7, no. 4, pp. 431–441, 2002.

[23] B. Salemi, M. Moll and W. Shen, 'Superbot: A deployable, multifunctional, and modular self-reconfigurable robotic system', in IEEE/RSJ International Conference on Intelligent Robots and Systems (IROS), pp. 3636–3641, 2006.

[24] D. Brandt, D. Christensen and H. Lund, 'Atron robots: Versatility from self-reconfigurable modules', in International Conference on Mechatronics and Automation, 2007. (ICMA), pp. 26–32, 2007.

[25] N. Ranasinghe, J. Everist and W.-M. Shen, 'Modular robot climbers', in Workshop on Self-Reconfigurable Robots, Systems & Applications in IEEE/RSJ International Conference on Intelligent Robots and Systems (IROS), 2007.

[26] F. Hou, N. Ranasinghe, B. Salemi and W. Shen, 'Wheeled locomotion for payload carrying with modular robot', IEEE/RSJ International Conference on Intelligent Robots and Systems (IROS), pp. 1331–1337, 2008.

[27] T. Matsumaru, 'Design and control of the modular robot system: Tomms', in IEEE International Conference on Robotics and Automation (ICRA), pp. 2125–2131, 1995.

[28] C. J. Paredis, H. B. Brown and P. K. Khosla, 'A rapidly deployable manipulator system', in IEEE International Conference on Robotics and Automation, pp. 1434–1439, 1996.

[29] J. Baca, M. Ferre and R. Aracil, 'A heterogeneous modular robotic design for fast response to a diversity of tasks', Robotics and Autonomous Systems, vol. 60, no. 4, pp. 522–531, 2012.

[30] Y. Guan, L. Jiang, X. Zhang, H. Zhang and X. Zhou, 'Development of novel robots with modular methodology', IEEE/RSJ International Conference on Intelligent Robots and Systems (IROS), pp. 2385–2390, 2009.

[31] Y. Guan, H. Zhu, W. Wu, X. Zhou, Li Jiang, C. Cai et al., 'A Modular Biped Wall-Climbing Robot With High Mobility and Manipulating Function', IEEE/ASME Transactions on Mechatronics, vol. 18, no. 6, 2013.

[32] M. Nilsson, 'Connectors for self-reconfiguring robots', IEEE/ASME Transactions on Mechatronics, vol. 7, no. 4, pp. 473–474, 2002.

[33] E. Guizzo and E. Ackerman, 'The rise of the robot worker', IEEE Spectrum, vol. 49, no. 10, pp. 34–41, 2012.

2

The Dynamic Characteristics of a Manipulator with Parallel Kinematic Structure Based on Experimental Data

S. Osadchy[1], V. Zozulya[2] and A. Timoshenko[3]

[1,2]Faculty of Automation and Energy, Kirovograd National Technical University, Ukraine
[3]Faculty of Automation and Energy, Kirovograd Flight Academy National Aviation University, Ukraine
Corresponding author: S. Osadchy <srg2005@ukr.net>

Abstract

The chapter presents two identification techniques which the authors found most useful in examining the dynamic characteristics of a manipulator with a parallel kinematic structure as an object of control. These techniques emphasize a frequency domain approach. If all input/output signals of an object can be measured, then the first one of such techniques may be used for identification. In the case when all disturbances can't be measured, the second identification technique may be used.

Keywords: Manipulator with parallel kinematics, structural identification, control system.

2.1 Introduction

Mechanisms with parallel kinematics [1, 2] compose the basis for the construction of single-stage and multi-stage manipulators. A single-stage manipulator consists of an immobile basis, a mobile platform and six guide rods. Each rod can be represented as two semi rods A_{ij} and an active kinematics pair B_{ij} (Figure 2.1).

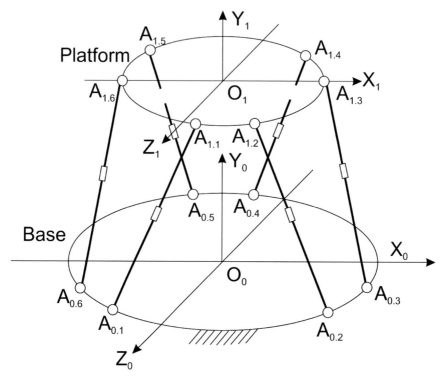

Figure 2.1   Kinematic diagram of single-section mechanism.

We will consider two systems of co-ordinates: inertial $O_0X_0Y_0Z_0$ with the origin in the center of the base O_0 and mobile $O_1X_1Y_1Z_1$, with the origin O_1 in the platform center of mass. From Figure 2.1 it is evident, that such mechanisms consist of thirteen mobile links and eighteen kinematics pairs. That is why, in accordance with [2], the number of its possible motions equals six.

Let us propose the following definitions ($j \in [1{:}6]$): l_{1j} – length of rod number j; M_x, M_y, M_z – projections of the net resistance moment vector on the axes of the co-ordinate system $O_0X_0Y_0Z_0$.

Obviously, while lengths $l_{1,j}$ are changing, then the co-ordinates of the platform's center of mass and the projections of the resistance moment vector are changing too.

From the point of view of automatic control theory, the mechanism with parallel kinematics belongs to the array of mobile control objects with two multidimensional entrances (control signals and disturbances) and one output vector (platform co-ordinates).

2.2 Purpose and Task of Research

The main purpose of this research is to construct a mathematical model which characterizes the interrelation between control signals, disturbances and co-ordinates of the platform center of mass on the base of experimental data.

If one assembles the length of rod changes in the vector of control signals u_1, the projections of force resistance moment changes in the disturbance vector ψ and the coordinates of the platform center of mass changes in the output vector x

$$u_1 = \begin{bmatrix} l_{1,1} \\ \vdots \\ l_{1,6} \end{bmatrix}, \psi = \begin{bmatrix} M_x \\ M_y \\ M_z \end{bmatrix}, x = \begin{bmatrix} X_c \\ Y_c \\ Z_c \end{bmatrix}, \tag{2.1}$$

then the block diagram of the mechanism with parallel kinematics can be represented as shown on the Figure 2.2 where W_u is an operator which characterizes the influence of the control signals vector u on the output vector x and W_ψ is an operator which describes the influence of the disturbance vector ψ on the output vector x. In this case, in order to find the mathematical model, it is necessary to define these operators. If we want to find such operators

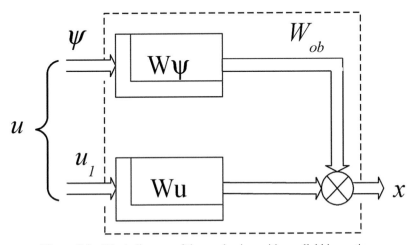

Figure 2.2   Block diagram of the mechanism with parallel kinematics.

based on experimental data, then two variants of the research task can be enunciated.

The first variant will be applied if the components of vectors u_1, x and ψ can be measured fully (the complete data). The second variant will be applied in the case when only the components of vectors u1 and x can be measured (the incomplete data).

So the research on the dynamics of the mechanism with parallel kinematics can be formulated as follows: to find transfer function matrices W_u, W_ψ and also to estimate the influence of vectors u_1 and ψ on vector x on the base of known complete or incomplete experimental data.

The solution of such a problem has been found as a result of three stages of:

- The development of algorithms for the structural identification of a multivariable Dynamic object with the help of complete or incomplete data;
- Collecting and processing experimental data about vectors u_1, x and ψ;
- The verification of the results of the structural identification.

2.3 Algorithm for the Structural Identification of the Multivariable Dynamic Object with the Help of the Complete Data

Let's suppose the identification object dynamics is characterized by a transfer function matrix W_{ob} (Figure 2.2), which may have unstable poles. Suppose that as a result of the processing of regular components of vectors *u* and *x,* the Laplace transformations $\widehat{U}_p$ and $\widehat{X}_p$ are defined

$$\widehat{U}_p = L\{u\} = L\left\{\begin{bmatrix} u_1 \\ \psi \end{bmatrix}\right\}, \widehat{X}_p = L\{x\}. \tag{2.2}$$

Thus, the Laplace transformation of output vector $\widehat{X}_p$ has unstable poles of vector $\widehat{U}_p$ and unstable poles of matrix W_{ob}. Therefore, it is possible to remove all unstable poles from $\widehat{X}_p[3]$ which differ from the unstable poles of $\widehat{U}_p$ and to define a diagonal polynomial matrix W_2 such as

$$\widehat{Y}_p = W_2 \cdot \widehat{X}_p . \tag{2.3}$$

In this case, the interdependence between vectors $\widehat{Y}_p$ and $\widehat{U}_p$ is expressed with the help of equation

$$\widehat{Y}_p = F_{1p} \cdot \widehat{U}_p, \qquad (2.4)$$

where F_{1p} is a transfer function matrix, all the poles of which are located in the left half-plane (LHP) of complex variables. It is equal to

$$F_{1p} = W_2 \cdot W_{ob}. \qquad (2.5)$$

Consequently, the identification problem consists in determining a physically implemented matrix F_{1p} that minimizes a quality functional

$$J = \frac{1}{2 \cdot \pi \cdot j} \cdot \int_{-j\infty}^{j\infty} tr(\varepsilon \cdot \varepsilon_* \cdot A) \cdot ds, \qquad (2.6)$$

where ε – identification error, which is equal to

$$\varepsilon = F_{1p} \cdot \widehat{U}_p - \widehat{Y}_p, \qquad (2.7)$$

$A-$ is a positively defined polynomial weight matrix.
To solve this problem, the ratio (2.7) must be submitted in a vector-matrix form

$$\varepsilon = \begin{bmatrix} F_{1p} & -E_n \end{bmatrix} \cdot \begin{bmatrix} \widehat{U}_p \\ \widehat{Y}_p \end{bmatrix}. \qquad (2.8)$$

The Hermitian conjugated vector ε_* from Equation (2.6) is equal to

$$\varepsilon_* = \begin{bmatrix} \widehat{U}_{p*} & \widehat{Y}_{p*} \end{bmatrix} \cdot \begin{bmatrix} F_{1p*} \\ -E_n \end{bmatrix}. \qquad (2.9)$$

After introducing the expressions (2.8) and (2.9) into the Equation (2.6), the quality functional can be shown as follows:

$$J = \frac{1}{2 \cdot \pi \cdot j} \cdot \int_{-j\infty}^{j\infty} tr \left\{ \begin{bmatrix} F_{1p} & -E_n \end{bmatrix} \times \right.$$

$$\left. \times \begin{bmatrix} \widehat{U}_p \cdot \widehat{U}_{p*} & \widehat{U}_p \cdot \widehat{Y}_{p*} \\ \widehat{Y}_p \cdot \widehat{U}_{p*} & \widehat{Y}_p \cdot \widehat{Y}_{p*} \end{bmatrix} \cdot \begin{bmatrix} F_{1p*} \\ -E_n \end{bmatrix} \cdot A \right\} \cdot ds. \qquad (2.10)$$

Thus, the problem of structural identification is reduced to the minimization of the functional (2.10) on the class of a steady variation matrix of transfer functions F_{1p}. Such minimization has been carried out as a result of the application of the Wiener-Kolmogorov procedure. In accordance with such procedure [5], the first variation of the quality functional (2.10) has been defined as

$$\delta J = \frac{1}{2 \cdot \pi \cdot j} \cdot \int tr\{A_{0*} \cdot [A_0 \cdot F_{1p} \cdot D - (H_0 + H_+ + H_-)] \cdot L \cdot D_*$$
$$\times \; \delta F_{1p*} + \delta F_{1p} \cdot D \cdot L \cdot [D_* \cdot F_{1p*} \cdot A_{0*} - (H_0 + H_+ + H_-)_*]$$
$$\times \; A_0\} \cdot ds, \tag{2.11}$$

where A_0 is a result of the factorization [4] of the matrix A the determinante pf which has zeros with the negative real parts

$$A = A_{0*} \cdot A_0; \tag{2.12}$$

D is a fraction-rational matrix with particularities in the left half-plane (LHP) which is defined on the basis of the algorithms in articles [3, 4] from the following equation

$$D \cdot L \cdot D_* = \widehat{U}_p \cdot \widehat{U}_{p*}, \tag{2.13}$$

where L is a singular matrix, each element of which is equal to one; bottom index * designates the Hermitian conjugate operation; $H_0 + H_+ + H_-$ is a fraction-rational matrix which is equal to

$$H_0 + H_+ + H_- = A_0 \cdot \widehat{Y}_p \cdot \widehat{U}_{p*} \cdot D_*^{-1} \cdot L^+; \tag{2.14}$$

L^+ is the pseudo inverse to matrix L [5]; matrix H_0 is the result of the division; H_+ is a proper fractional rational matrix with poles that are analytic only in the right half-plane (RHP); H_- is a proper fractional rational matrix with poles that are analytic in LHP. In accordance with the chosen minimization procedure, a steady and physically realized variation F_{1p} which delivers a minimum to the functional (2.10) is equal to

$$F_{1p} = A_0^{-1} \cdot (H_0 + H_+) \cdot D^{-1}. \tag{2.15}$$

If one takes into account matrices W_2, F_{1p} from Equations (2.3) and (2.15), then an unknown object transfer function matrix W_{ob} can be identified with the help of the following expression

$$W_{ob} = W_2^{-1} \cdot F_{1p}. \tag{2.16}$$

The separation [4] of the transfer function matrix (2.16) makes it possible to find the unstable part of the object transfer function matrix with the help of equation

$$W_{ob2} = W_-, \tag{2.17}$$

where W_- is a fraction-rational matrix with particularities in the RHP.

An algorithm for the structural identification of the multivariable dynamic object with an unstable part on the base of the vectors u and x emplees the implementation of the following operations:

- Search the matrix W_2 as a result of the left-hand removal of the unstable poles from X_p, which differ from the poles of U_p;
- Factorization of the weight matrix A from (2.12);
- Identification of the analytical complex variable matrix D Equation (2.13);
- Calculation of $H_0 + H_+$ as a result of the separation (2.14);
- Calculation of F_{1p} on the basis of the Equation (2.15);
- Identifying W_{ob2} by the separation of the product (2.16).

In this way, we have substantiated the algorithm for the structural identification of the multivariable dynamic object with the help of the complete experimental data.

2.4 Algorithm for the Structural Identification of the Multivariable Dynamic Object with the Help of Incomplete Data

Let's suppose that the identification object dynamics is characterized by a system of ordinary differential equations with constant coefficients. The Fourier transformation of such system, subject to the zero initial conditions, can be shown as follows:

$$P \cdot x = M \cdot u_1 + \psi, \tag{2.18}$$

where P and M are polynomial matrices of the appropriate dimensions; ψ is the Fourier image of a centered multivariable stationary random process with the unknown spectral densities matrix $S_{\psi\psi}$. Let us admit also that vectors u and x are the centered multivariable stationary random processes with the matrices of the spectral and cross-spectral densities S_{xx}, S_{uu}, S_{xu}, S_{ux} known as a result of the experimental data processing. It is considered that the random process ψ can be formed by a filter with the transfer function matrix Ψ and is equal to

$$\psi = \Psi \cdot \Delta, \tag{2.19}$$

where Δ is the vector of the single δ-correlated "white" noises.

If one takes into account expression (2.19), then Equation (2.18) can be rewritten as follows:

$$x = P^{-1} \cdot M \cdot u_1 + P^{-1} \cdot \Psi \cdot \Delta \tag{2.20}$$

and a transfer function matrix which must be identified can be defined as the expression

$$\phi = \begin{bmatrix} \phi_{11} & \phi_{12} \end{bmatrix} = \begin{bmatrix} P^{-1} \cdot M & P^{-1}\Psi \end{bmatrix}. \tag{2.21}$$

So, the Equation (2.20) can be simplified to the form

$$x = \phi \cdot y, \tag{2.22}$$

where y is an extended vector of the external influences

$$y = \begin{pmatrix} u' & \Delta' \end{pmatrix}'. \tag{2.23}$$

Thus, the identification problem can be formulated as follows. Using the records of the vectors x and y, choose the sectional matrix string ϕ (2.21) that provides minimum to the following quality functional

$$J = \frac{1}{j} \int\limits_{-j\infty}^{j\infty} tr\{S'_{\varepsilon\varepsilon}R\}ds, \tag{2.24}$$

where J is equal to the sum of the identification errors variances as the elements of the identification errors vector ε

$$\varepsilon = x - \phi \cdot y. \tag{2.25}$$

$S'_{\varepsilon\varepsilon}$ is a transposed matrix of the identification errors spectral densities

$$S'_{\varepsilon\varepsilon} = S'_{xx} - S'_{yx}\phi_* - \phi S'_{xy} + \phi S'_{yy}\phi_*; \tag{2.26}$$

$$S'_{yx} = \begin{pmatrix} S'_{ux} & S'_{\Delta x} \end{pmatrix}; \tag{2.27}$$

$$S'_{yy} = \begin{bmatrix} S'_{uu} & O_{m\times n} \\ O_{n\times m} & S'_{\Delta\Delta} \end{bmatrix}, \tag{2.28}$$

S'_{xx}, S'_{uu}, S'_{yy} are the transposed spectral density matrices of the vectors x, u, y; S'_{xy}, S'_{yx}, S'_{ux} is the transposed cross-spectral density matrices between vectors x and y, y and x, u and x; $S'_{\Delta x}$ is a transposed cross-spectral density matrix which is found on the basis of the Wiener's factorization [3] of the additional connection equation

$$S_{x\Delta}S_{\Delta x} = S_{xx} - S_{xu}S_{uu}^{-1}S_{ux}, \qquad (2.29)$$

R is a positively defined polynomial weight matrix.

An algorithm for the set problem decision, which is grounded in [8] and allows defining the sought after matrix ϕ which minimizes the functional (2.24), has the following form

$$\phi = R_0(K_0 + K_+)D^{-1}, \qquad (2.30)$$

in which matricees R_0 and D are results of the Wiener's factorization [3] of matrices R and S'_{yy} so that

$$R_{0*}R_0 = R; DD_* = S'_{yy}; \qquad (2.31)$$

$K_0 + K_+$ is a transfer function matrix with the stable poles, which is defined as a result of the following equation right part separation [7]

$$K_0 + K_+ + K_- = R_0 S'_{yx} D_*^{-1}. \qquad (2.32)$$

An algorithm for the structural identification of a multivariable dynamic object with the help of the stochastic stationary components of vectors u_1 and x implies the following operations:

- Search for the spectral and cross-spectral densities matrices S_{xx}, S_{uu}, S_{yy}, S_{ux}, S_{xu} on the base of the experimental data processing;
- Factorization of the weight matrix R from (2.31);
- Factorization of the additional connection Equation (2.29);
- Factorization of the transposed spectral densities matrix (2.28);
- The Wiener's separation of the matrix (2.32);
- Calculation of matrix ϕ based on Equation (2.30);
- Identification of matrices ϕ_{11} and ϕ_{12} with the help of Equation (2.21).

In this way, we substantiate the algorithm for the structural identification of the multivariable object on the base of the incomplete experimental data.

2.5 The Dynamics of the Mechanism with a Parallel Structure Obtained by Means of the Complete Data Identification

To identify the models of a dynamics, we used tracks of changes in the components of vectors u, ψ and x, obtained as a result of the behavior of a modeling platform using a virtual model. The case was considered when the motion platform center of mass O_1 remained in the plane of the manipulator symmetry $O_0X_0Y_0$. Thus, it is evident that in this case, instead of six rods only three (Figure 2.1) may be considered and the dimension of vector u (2.1) is equal 3.

As a result of the computational experiment, all the above vectors' components were obtained and all graphs of their changes (Figures 2.3–2.5) were built.

For solving of the identification problem, the control and perturbations vectors were combined into a single vector u of the input signals

$$u = \begin{bmatrix} l_{1,1} & l_{1,2} & l_{1,3} & M_x & M_y & M_z \end{bmatrix}^T. \qquad (2.33)$$

In this case, the identification problem is to estimate the order and the parameters of a differential equations system which characterizes the mechanism motion.

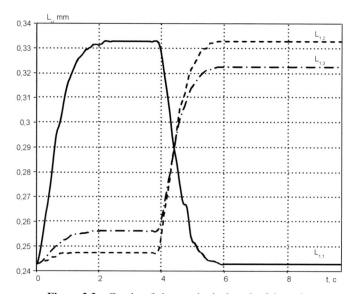

Figure 2.3   Graphs of changes in the length of the rods.

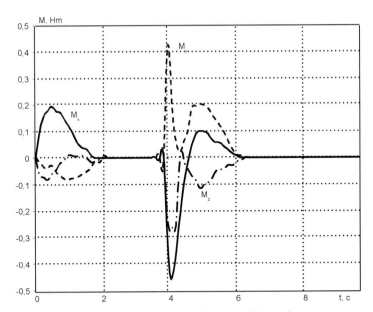

Figure 2.4 Graphs of changes in the projections of the resistance moments.

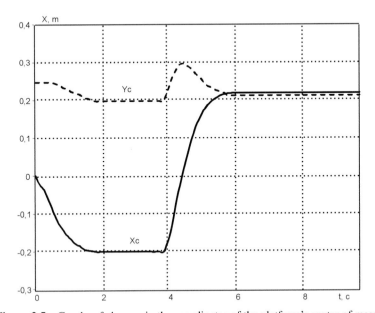

Figure 2.5 Graphs of chances in the coordinates of the platform's center of mass.

The state space dynamic model of the mechanism *is defined with the help of the* System Identification Toolbox of the Matlab environment. Considering the structure of vector u, defined by (2.33), allows to obtain the equation of the hexapod's state as follows:

$$\dot{y}(t) = Ay(t) + B_u u(t) + B_\psi \psi(t),$$

$$x(t) = Cy(t) + D_u u(t) + D_\psi \psi(t), \tag{2.34}$$

where the matrices B_u, B_ψ, D_u and D_ψ are easily determined.

The analysis of the obtained model of dynamics shows that the moving object is fully controllable and observable.

As a result, the Laplace transformation of the left and right parts of the Equations (2.34) obtained the following relations

$$\begin{cases} (sE_n - A)y(s) = B_u u(s) + B_\psi \psi(s) \\ x(s) = Cy(s) + D_u u(s) + D_\psi \psi(s), \end{cases} \tag{2.35}$$

where $y(s)$, $x(s)$, $u(s)$, $\psi(s)$ – the Laplace image of the vector, E_n – the identity matrix, s – the independent complex variable.

After solving the system of Equations (2.35) with respect to the vector of the initial coordinates of the mechanism x, the following matrices of the transfer functions W_u and W_ψ (Figure 2.2) were obtained

$$W_u = C (sE_n - A)^{-1} B_u + D_u, \tag{2.36}$$

$$W_\psi = C (sE_n - A)^{-1} B_\psi + D_\psi. \tag{2.37}$$

Substituting the appropriate numerical matrices C, B_u, B_ψ, D_u, D_ψ, in expressions (2.36), (2.37) allowed determining that

$$W_u = \left[\begin{array}{cc} \dfrac{-0.37(s+1.606)(s^2+21.03s+790.2)}{(s+1.55)(s^2+3.463s+117.8)} & \dfrac{-0.599(s-57.51)(s^2-2.215s+8.456)}{(s+1.55)(s^2+3.463s+117.8)} \\ \dfrac{0.29(s+2.232)(s^2+21.65s+529.7)}{(s+1.55)(s^2+3.463s+117.8)} & \dfrac{2.23(s-14.01)(s^2-0.67s+87.82)}{(s+1.55)(s^2+3.463s+117.8)} \end{array} \right.$$

$$\left. \begin{array}{c} \dfrac{0.92796(s+0.4741)(s^2-8.066s+404.6)}{(s+1.55)(s^2+3.463s+117.8)} \\ \dfrac{0.36433(s+75.92)(s^2-2.029s+91.39)}{(s+1.55)(s^2+3.463s+117.8)} \end{array} \right], \tag{2.38}$$

$$W_\psi = \begin{bmatrix} \dfrac{-0.006(s-2.782)(s^2+10.9s+450.5)}{(s+1.55)(s^2+3.463s+117.8)} & \dfrac{0.0023(s+43.94)(s^2-3.357s+8.447)}{(s+1.55)(s^2+3.463s+117.8)} \\[2mm] \dfrac{0.046892(s-10.1)(s^2+3.372s+133.1)}{(s+1.55)(s^2+3.463s+117.8)} & \dfrac{0.059(s+0.162)(s^2-2.875s+124.3)}{(s+1.55)(s^2+3.463s+117.8)} \end{bmatrix}$$

$$\begin{bmatrix} \dfrac{0.0049(s+57.79)(s+8.592)(s-2.702)}{(s+1.55)(s^2+3.463s+117.8)} \\[2mm] \dfrac{-0.014(s+22.92)(s^2+1.572s+174.3)}{(s+1.55)(s^2+3.463s+117.8)} \end{bmatrix}. \quad (2.39)$$

The analysis of the matrix structure of (2.38) and (2.39) and the Bode diagrams (Figure 2.6) shows that this mechanism can be classified as a multi-resistant mechanical filter with the input signals and the disturbances energy bands lying in the filter spectral band pass. The eigen frequency of such a filter is close to $11s^{-1}$ and depends on the moments of inertia and the mass of the moving elements of the mechanism (Figure 2.1).

The ordinary differential equations dynamics model of the system can be obtained if you present the transfer function matrices W_u and W_ψ as a product of the polynomial matrices P, M and M_ψ with the minimum order polynomials:

$$W_u = P^{-1}M, \quad (2.40)$$

$$W_\psi = P^{-1}M_\psi. \quad (2.41)$$

To find the polynomial matrices P and M with the minimum order polynomials, we propose the following algorithm:

- By moving the poles to the right [3], the transfer function matrix W_u should be introduced as follows:

$$W_u = N_R D_R^{-1} \quad (2.42)$$

and the diagonal polynomial matrix D_R should be found;

- On the basis of the polynomial matrices N_R and D_R found by the CMFR algorithm, substantiated in [7], the unknown matrices P and M should be identified

$$P^{-1}M = N_R D_R^{-1}; \quad (2.43)$$

- From Equation (2.41) and the known matrices P and W_ψ the polynomial matrix M_ψ should be found

$$M_\psi = PW_\psi. \quad (2.44)$$

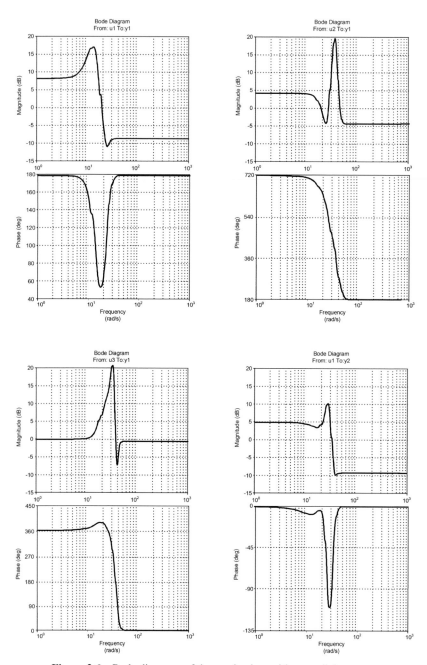

Figure 2.6   Bode diagrams of the mechanism with a parallel structure.

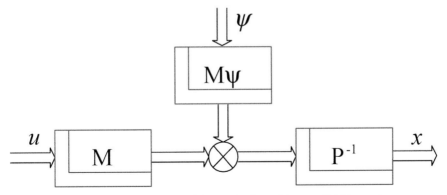

Figure 2.7 Block diagram of the mechanism with a parallel kinematics.

The application of the algorithms (2.42) and (2.43) to the original data represented by expressions (2.38), (2.39) made it possible to obtain the polynomial matrices P, M, M_ψ. Then, application of the inverse Laplace transform under the zero initial conditions allowed determining the following system of ordinary differential equations

$$P_0\ddot{x} + P_1\dot{x} + P_2 x = M_0\ddot{u} + M_1\dot{u} + M_2 u + M_{\psi 0}\ddot{\psi} + M_{\psi 1}\dot{\psi} + M_{\psi 2}\psi, \quad (2.45)$$

where P_i, M_i, ψ_i – the numeric matrices.

Representing the dynamic model (2.45) allowed to reconstruct a block diagram of a parallel kinematics mechanism in the standard form (Figure 2.7), where P^{-1} is an inverse of matrix P.

2.6 The Dynamics of the Mechanism with a Parallel Structure Obtained by Means of the Incomplete Data Identification

The incomplete experimental data arises when not all entrance signals of vector u (Figure 2.2) can be measured and recorded. Such a situation appears during the dynamics identification of a manipulator with controlled diode motor-operated drive (Figure 2.8). In this case, only signals of the platforms center of mass set position which form vector u_1 and signals of the platform's center of mass current position which form vector x are accessible for measuring. Thus, the task of the identification is to define matrices ϕ_{11}, ϕ_{12} from Equation (2.21) by records of vectors u_1 and x.

Figure 2.8   Manipulator with a controlled diode motor-operated drive.

Solving this task is achieved as a result of algorithm (2.30–2.32) applied to the estimations of the spectral and cross-spectral density matrices S_{uu}, S_{xx}, S_{ux} and $S_{\Delta x}$. For the illustration of this algorithm application, we used the records of the «input-output» vectors u, x with the following structure

$$u = \begin{bmatrix} u_1 \\ u_2 \\ u_3 \end{bmatrix} ; x = [Z_c] , \tag{2.46}$$

where u_1 – the required value of the manipulators' platform center of mass O_1 projection on the axis O_0X_0 (Figure 2.1); u_2 – the required value of the manipulators platform center of mass O_1 projection on the axis O_0Y_0; u_3 – the required value of the manipulators' platform center of mass O_1 projection on the axis O_0Z_0 (Figure 2.1).

As a result of the experiment, all the above vectors' (2.46) components were obtained and all the graphs of their changes (Figures 2.9, 2.10) were built.

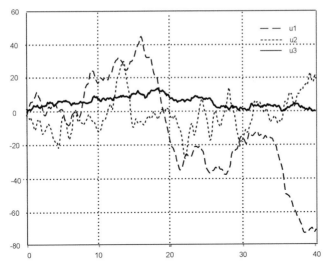

Figure 2.9   Grapas of the vector u componente changes.

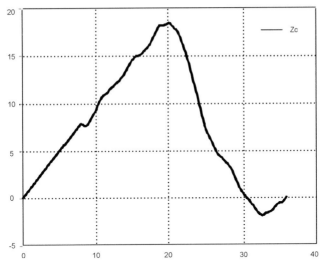

Figure 2.10   Grapas of vector x component changes.

In accordance with this algorithm (2.26–2.32) on the first stage of the calculations, estimations of matrices S_{uu}, S_{xx}, S_{ux} were found.

Approximation of such estimations made it possible to construct the following spectral densities matrices with the help of the logarithmic frequency descriptions method [4]

$$S'_{uu} = \begin{bmatrix} \frac{7.87|s+0.12|^2}{|s^2+0.29s+0.034|^2} & 0 & \frac{12(s+0.095)(-s+0.15)}{|s^2+0.29s+0.034|^2} \\ 0 & \frac{45.5|s+0.075|^2}{|s^2+0.64s+0.16|^2} & 0 \\ \frac{12(-s+0.095)(s+0.15)}{|s^2+0.29s+0.034|^2} & 0 & \frac{2.88|s+0.095|^2}{|s^2+0.29s+0.034|^2} \end{bmatrix};$$

$$S'_{xx} = \frac{4.59\,|(s+0.08)\,(s+1.3)|^2}{|(s+0.8)\,(s^2+0.29s+0.034)|^2};$$

$$S'_{ux} = \begin{bmatrix} \frac{4.89(-s+0.1)(s+0.2)(-s+2.1)}{(s+0.8)|s^2+0.29s+0.034|^2} & 0 \end{bmatrix}$$

$$\frac{1.2\,(s+3)\,|s+0.095|^2}{(s+0.8)|s^2+0.29s+0.034|^2}\Bigg]. \tag{2.47}$$

The introduction of the found matrices (2.47) into the Equation (2.29) and its factorization made it possible to find the cross spectral density $S_{\Delta x}$

$$S_{\Delta x} = \frac{0.89\,(s+1.6)\,(s+0.031)}{(s+0.8)\,(s^2+0.29s+0.034)}. \tag{2.48}$$

The factorization [4] of the transposed spectral densities matrix S'_{yy} from expression (2.31) allowed finding the following matrix D

$$D = \begin{bmatrix} \frac{8.87(s+0.1)}{s^2+0.29s+0.034} & 0 & \frac{-0.54}{s^2+0.29s+0.034} & 0 \\ 0 & \frac{6.75(s+0.075)}{s^2+0.64s+0.16} & 0 & 0 \\ \frac{1.34(s+0.11)}{s^2+0.29s+0.034} & 0 & \frac{1.04(s-0.057)}{s^2+0.29s+0.034} & 0 \\ 0 & 0 & 0 & 1 \end{bmatrix}. \tag{2.49}$$

Taking into account the dimension of the output co-ordinates vector x (2.46), we have accepted that matrix R is equal to the identity matrix. At that rate matrix R_0 also equals 1. Substitution of the results (2.47–2.49) in expression (2.32) and its separation allowed defining that

$$K_0 + K_+ = \begin{bmatrix} \frac{-0.44(s-2.57)(s+0.12)}{(s+0.8)(s^2+0.29s+0.034)} \\ 0 \\ \frac{1.95(s+1.16)(s-0.04)}{(s+0.8)(s^2+0.29s+0.034)} \\ \frac{0.89(s+1.6)(s+0.03)}{(s+0.8)(s^2+0.29s+0.034)} \end{bmatrix}'. \tag{2.50}$$

Substitution of matrices (2.49), (2.50) in expression (2.30) and taking into account the vectors u and x made it possible to solve the problem and to find such matrices ϕ_{11} and ϕ_{12} as

$$\phi_{11} = \begin{bmatrix} \frac{-0.33(s+0.61)(s-0.034)}{(s+0.03)(s+0.8)} \\ 0 \\ \frac{1.87(s+1.12)(s-0.008)}{(s+0.03)(s+0.8)} \end{bmatrix}' ; \qquad (2.51)$$

$$\phi_{12} = \frac{0.89\,(s+1.6)\,(s+0.031)}{(s+0.8)\,(s^2+0.29s+0.034)} \qquad (2.52)$$

Taking into account the flow diagram on Figure 2.2 and the physical sense of matrices ϕ_{11} and ϕ_{12} made it possible to formulate the equation

$$W_u = \phi_{11}; \quad W_\psi = \phi_{12}. \qquad (2.53)$$

For the definition of the incomplete data identification error, the Equation (2.26) is used and the error spectral density is found in the form which is shown below:

$$S'_{\varepsilon\varepsilon} = \frac{0.023}{|s+0.037|^2}. \qquad (2.54)$$

The identification error mathematical mean is equal to zero and its relative variances is equal to

$$E_\varepsilon = \frac{\int\limits_{-j\infty}^{j\infty} S_{\varepsilon\varepsilon}ds}{\int\limits_{-j\infty}^{j\infty} S_{xx}ds} = 0.0157. \qquad (2.55)$$

Obviously it is clear that the main part of the error ε oscillations power density is concentrated in the area of the infrasonic frequencies. The presence of such an error is explained by the limited duration of the experiment.

2.7 Verification of the Structural Identification Results

Verification of the identification results was implemented with the help of the modeling tool SIMULINK from Matlab. The *principle* of the verification of the identification results exactness was based on the comparison of vector x records to the results of the platform center of mass position simulation.

On Figure 2.11, the simulation model block u is designed to create the set of the input signals. The block m generates a set of projections of the net

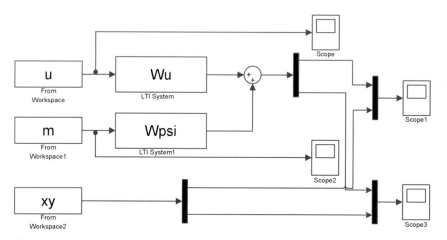

Figure 2.11   The scheme simulation model of the mechanism with parallel kinematics.

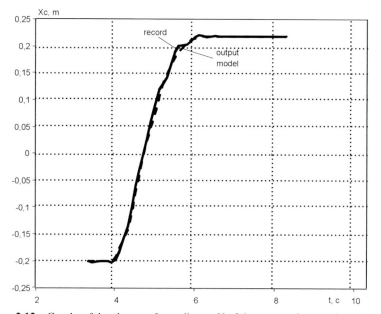

Figure 2.12   Graphs of the change of coordinates X of the center of mass of the platform.

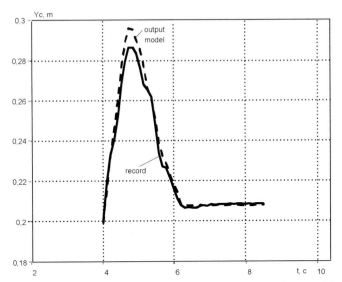

Figure 2.13 Graphs of thee changes of Y coordinates of the center of mass of the platform.

resistance moment vector on the axis of the co-ordinate system $O_0X_0Y_0Z_0$. The output of the block xy formed the vector x of records. The blocks Wu and Wpsi are designed for storing the matrices of the transfer functions W_u and W_ψ.

According to the simulation results, the grapas (Figures 2.12 and 2.13) are built. Analysis of these graphs show that they are close enough.

2.8 Conclusions

The conducted research on the mechanism with a parallel structure dynamics made it possible to obtain the following scientific and practical results:

- Substantiate two new algorithms for the structural identification of the multivariable moving object dynamic models. The first one of them allows to define the structure and parameters of a transfer function matrix of the object with unstable poles using the regular vectors "input-output". The second one allows identifying not only the model of a mobile object but also the model of the non-observed stationary stochastic disturbance;
- Three types of models which characterize the dynamics of the manipulator with a parallel kinematics are identified. This allows to use different modern multidimensional optimal control systems synthesis methods for designing the optimal mechatronic system;

- It is shown that the mechanism with a parallel kinematics as an object of control is a multi-resistant mechanical filter.

References

[1] J. P. Merlet, 'Parallel robots. Solid mechanics and its application', V.74 – Kluwer Academic Publishers, 2000, pp. 394

[2] S. V. Volkomorov, J. P. Hagan, A. P. Karpenko, 'Modeling and optimization of some parallel mechanisms', Information Technology. Application 2010, No. 5. pp.1–32 (in Russian).

[3] M. C. Davis, 'Factoring the spectral matrix', IEEE Trans. Automat. Cointr. – 1963.- AC-8, N 4, pp. 296–305.

[4] V. N. Azarskov, L. N. Blokhin, L. S. Zhitetsky, 'The methodology of constructing optimal systems stochastic stabilization', Monograph, Ed. Blokhin L. N. - K.: NAU Book Publishers, 2006, pp. 440 - Bibliography: pp. 416–428 (in Russian).

[5] F. R. Gantmakher, 'Theory matrits.-4th ed', Nauka, 1988. p. 552 (in Russian).

[6] V. Kucera 'Discrete line control: the polynomial equation approach'. Praha: Akademia, 1979, pp. 206

[7] F. A. Aliev, V. A. Bordyug, V. B. Larin, 'Time and frequency methods for the synthesis of optimal regulators', Baku: Institute of Physics of the Academy of Sciences, 1988, pp. 46 (in Russian).

3

An Autonomous Scale Ship Model for Parametric Rolling Towing Tank Testing

M. Míguez González, A. Deibe, F. Orjales, B. Priego and F. López Peña

Integrated Group for Engineering Research, University of A Coruña, Spain
Corresponding author: M. Míguez González <mmiguez@udc.es>

Abstract

This chapter presents the work carried out for developing a self-propelled scale ship model for model testing, with the main characteristic of not having any material link to a towing device to carry out the tests. This model has been fully instrumented in order to acquire all the significant raw data, process them onboard and communicate with an inshore station.

Keywords: Ship scale model, autonomous systems, data acquisition, parametric resonance, towing tank tests.

3.1 Introduction

Ship scale model testing has traditionally been the only way to accurately determine ship resistance, propulsion, maneuvering and seakeeping characteristics. These tests are usually carried out in complex facilities, where a large carriage, to which the model is attached, moves it following a desired path along a water tank.

Ship model testing could be broadly divided into four main types of tests. Resistance tests, either in waves or still water, intended to obtain the resistance of the ship without taking the effects of its propulsion system into consideration. Propulsion tests aimed at analyzing the performance of the ship propeller when it is in operation together with the ship itself. Maneuvering tests the objective of which is to analyze the capability of the ship for carrying

Advances in Intelligent Robotics and Collaborative Automation, 49–72.

out a set of defined maneuvers. And finally, seakeeping tests aimed at studying the behavior of the ship while sailing in waves [1].

The former two tests are carried out in towing tanks, which are slender water channels where the model is attached to the carriage that tows it along the center of the tank. The latter two, on the other hand, are usually performed in the so-called ocean basins where the scale model can be either attached to a carriage or radio controlled with no mechanical connection to it.

However, there exist certain kinds of phenomena in the field of seakeeping that can be studied in towing tanks (which are cheaper and have more availability than ocean basins) and that are characterized by showing very large amplitude nonlinear motions. The interactions between the carriage and the model due to these motions, which are usually limited to a maximum in most towing tanks, and the lack of space under the carriage, reduce the applicability of slender channels for these types of tests.

One of these phenomena is that of ship parametric roll resonance, also known as parametric rolling. This is a well-known dynamical issue affecting ships, especially containerships, fishing vessels and cruise ships, and it can generate very large amplitude roll motions in a very sudden way, reaching the largest amplitudes in just a few rolling cycles. Parametric roll is due to the periodic alternation of wave crests and troughs along the ship, which produce the changes in ship transverse stability that lead to the aforementioned roll motions.

Resonance is most likely to happen when the ship sails in longitudinal seas and when a certain set of conditions are present, which include a wave encounter frequency ranging twice the ship's natural roll frequency, a wavelength almost equal to the ship length and a wave amplitude larger than a given threshold [2].

Traditionally, parametric roll tests in towing tanks have been carried out by using a carriage towed model, where the model is free to move in just some of the 6 degrees of freedom (typically heave, roll and pitch, the ones most heavily influencing the phenomenon) [3, 4]. However, this arrangement limits the possibility of analyzing the influence of the restrained degrees of freedom [5], which may also be of interest while analyzing parametric roll resonance, or it may interfere on its development.

The main objective of the present work is to overcome the described difficulties for carrying out scale tests in slender towing tanks where large amplitude motions are involved, while preserving the model capabilities for being used in propulsion (free-running), maneuvering and seakeeping tests.

This has been done by using a self-propelled and self-controlled scale model, able to freely move in the six degrees of freedom and to measure, store and process all the necessary data without direct need of the towing carriage. In addition, the model could be used for any other tests, both in towing tanks or ocean basins, with the advantage of being independent of the aforementioned carriage.

The type and amount of data to be collected is defined after taking into consideration the typology of tests to be carried out. Taking into account that all three, propulsion, maneuvering and seakeeping tests should be considered, all the data related with the motion of the ship, together with those monitoring the propulsion system and the heading control system, have to be collected. Moreover, a processing, storage and communication with shore system was also implemented and installed onboard.

Finally, the results obtained by using the model in a test campaign aimed at predicting the appearance of parametric roll, where a detection and prevention algorithm has been implemented in the onboard control system, are presented.

3.2 System Architecture

The first implementation of the proposed system has been on a scale ship model that has been machined from high-density (250 kg/m^3) polyurethane blocks to a scale of 1/15th, painted and internally covered by epoxy reinforced fiberglass. Mobile lead weights have been installed on supporting elements, allowing longitudinal, transverse and vertical displacements for a correct mass arrangement. Moreover, two small weights have been fitted into a transverse slider for fast and fine-tuning of both the transverse position of the center of gravity and the longitudinal moment of inertia.

The propulsion system consists of a 650W brushless, outrunner, three-phase motor, an electronic speed control (ESC) and a two-stage planetary gearbox, which move a four-bladed propeller. The ESC electronically generates a three-phase low voltage electric power source to the brushless motor. With this ESC, the speed and direction of the brushless motor can be controlled. It can even be forced to work as a dynamic brake if need be. The rudder is linked to a radio control servo with a direct push-pull linkage, so that the rudder deflection angle can be directly controlled with the servo command.

Both the brushless motor and the rudder may be controlled either by using a radio link or by the onboard model control system, which is the

Table 3.1   Measured Parameters

Type of Test	ITTC Guideline	Measured Parameters	
Propulsion	7.5-02-03-01.1	Model speed	√
		Resistance/External tow force	√
		Propeller thrust and torque	√
		Propeller rate of revolution	√
		Model trim and sinkage	√
Maneuverability (additional to Propulsion)	7.5-02-06-01	Time	√
		Position	√
		Heading	√
		Yaw rate	√
		Roll angle	√
		Rudder angle	√
		Force and moment on steering devices	x
Seakeeping (additional to Maneuverability)	7.5-02-07-02.1	Model motions (6 d.o.f.)	√
		Model motion rates (6 d.o.f.)	√
		Model accelerations (6 d.o.f.)	√
		Impact pressures	x
		Water on deck	x
		Drift angle	x

default behavior. The user can choose between both control methods any time; there's always a human at the external transmitter to ensure maximum safety during tests.

3.2.1 Data Acquisition

In order to ensure that the model can be used for propulsion, maneuvering and seakeeping tests and is in accordance with the ITTC (International Towing Tank Conference) requirements, the following parameters have been measured (Table 3.1):

In order to obtain data for all the representative parameters, the following sensors have been installed onboard:

- Inertial Measurement Unit (IMU): having nine MEMS embedded sensors; including 3 axis accelerometers, 3 axis gyros and 3 axis magnetometers. The IMU has an internal processor that provides information on accelerations in the OX, OY and OZ axis, angular rates around these three axes and quaternion based orientation vectors (roll, pitch, yaw), both in RAW format and filtered by using Kalman techniques. This sensor has been placed near the ship's center of gravity with the objective of improving its performance;

- Thrust sensor: a thrust gauge has been installed to measure the thrust generated by the propeller at the thrust bearing;
- Revolution and torque sensor: in order to measure the propeller revolutions and the torque generated by the engine, a torque and rpm sensor has been installed between both elements;
- Sonars: intended to measure the distance to the towing tank walls and feed an automatic heading control system;
- Not directly devoted to tested magnitudes, there are also a temperature sensor, battery voltage sensors and current sensors.

Data acquisition is achieved through an onboard mounted PC, placed forward on the bottom of the model. The software in charge of the data acquisition and processing and ship control is written in Microsoft .Net, and installed in this PC. This software is described in the following section. There is a Wi-Fi antenna at the ship's bow, connected to the onboard PC that enables a Wi-Fi link to an external, inshore workstation. This workstation is used to monitor the ship operation.

An overview of the model where its main components are shown is included in Figure 3.1.

3.2.2 Software Systems

In this section, the software designed and implemented to control the ship is described. Regarding the operational part of the developed software, there are three main tasks that have to be performed in real time in order to monitor

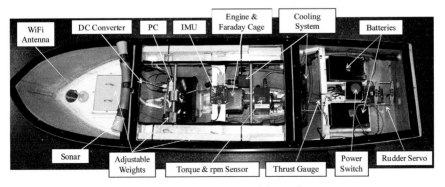

Figure 3.1   Ship scale model overview.

and control the ship: acquisition, computation and actuation. In every time step, once the system is working, all the sensor measurements are collected. The indispensable sensors that need to be connected to the system are: the Inertial Measurement Unit (IMU) which provides the acceleration, magnetic, angular rate and attitude measurements, the sonars and thrust gauge, in this case connected to a data acquisition board, and the revolution and torque sensor (Figure 3.2).

Once the sensor data are acquired, the system computes the proper signals to modify the rudder servo and the motor ESC commands. These signals can be set manually (using the software interface from the Wi-Fi linked workstation, or the external RC transmitter) or automatically. Applying a controller over the rudder servo, using the information from the sonar signals, it is possible to keep the model centered and in course along the towing tank. Another controller algorithm is used to control the ship speed.

The software is based on VB. NET and to interact with the system from an external workstation, a simple graphical user interface has been implemented (Figure 3.3).

From an external workstation, the user can start and finish the test, activate the sensors to measure, monitor the data sensors in real time, control the rudder servo and the motor manually using a slider or stop the motor and finish the test. All the acquired data measurements are saved in a file for a future analysis of the test.

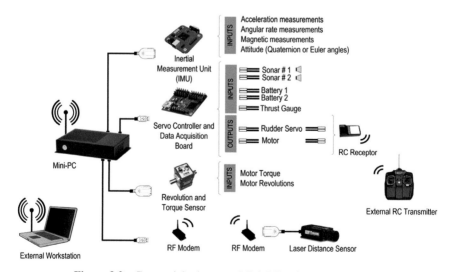

Figure 3.2   Connectivity between Mini-PC and sensors onboard.

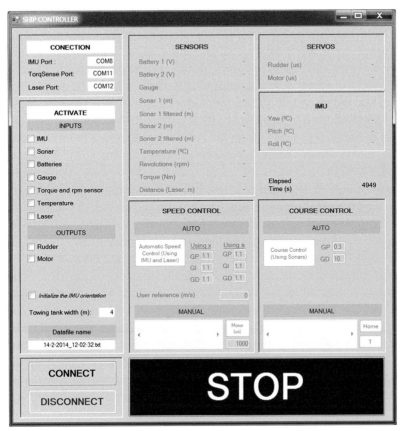

Figure 3.3 Graphical user interface to monitor/control the model from an external workstation.

3.2.3 Speed Control

The speed control of the model is done by setting an rpm command, which keeps engine revolutions constant by using a dedicated controller programmed in the governing software. Alternatively, a servo command may be used for setting a constant power input for the engine. In calm waters and for a given configuration of the ship model, there is a relationship between ship speed and propeller revolutions. By performing some preliminary tests at different speeds, this relation can be adjusted and used thereafter for testing within a simple, open loop controller. However, in case of testing with waves, these waves introduce an additional and strong drag component on the ship forward movement, and there is no practical way of establishing a similar sort of relationship. For these cases, the towing carriage is used as a reference and the speed is maintained by keeping the ship model in a steady relative position to the carriage.

The speed control strategy to cope with this composed speed was initially tested as shown in Figure 3.4. It was initially done by means of a double PID controller; the upper section of the controller tries to match the ship speed with a set point selected by the user, c_v. This portion of the controller uses the derivative of the ship position along the tank, x, as an estimation of the ship speed, e_{vx}. The position x is measured through the Laser Range Finder sensor placed at the beginning of the towing tank, facing the ship poop, to capture the ship position along the tank, and send this information through a dedicated RF modem pair, to the Mini-PC onboard. The bottom section, on the other hand, uses the integral of the ship acceleration in its local x-axis from the onboard IMU, v_a, as an estimation of the ship speed, e_{va}. Each branch has its own PID controller, and the sum of both outputs is used to command the motor.

Both speed estimations come from different sensors, in different coordinate systems, with different noise perturbations and, over all, they have different natures. The estimation based on the derivative of the position along the tank has little or zero drift over time, and its mean value matches the real speed on the tank x axis, and changes slowly. On the other hand, the estimation based on the acceleration along the ship's local x-axis is computed by the onboard IMU, from its MEMS sensors, and is prone to severe noises, drift over time and changes quickly. Furthermore, the former estimation catches the slow

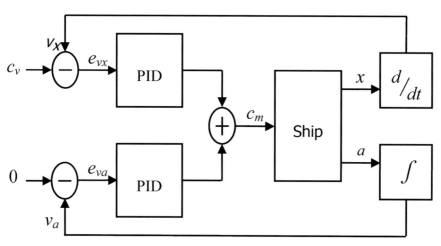

Figure 3.4   Double PID Speed control.

behavior of the ship speed, and the latter its quick changes. This is the reason to use different PID controllers with both estimations. The resulting controller follows the user-selected speed setpoint, with the upper branch eliminating any steady-state speed error, while minimizing quick speed changes with the lower branch.

Later on, a different speed control approach was introduced in order to improve its performance. Being the Laser Range Finder output an absolute measure of ship position, the speed obtained from its derivative is significantly more robust than the one estimated from the IMU in the first control scheme, and has no drift over time or with temperature. The new speed control algorithm is based in a complementary filter [6]. This filter estimates the ship speed from two different speed estimations, with different and complementary frequency components, as shown in Figure 3.5.

This two signal complementary filtering is based upon the use and availability of two independent noisy measurements of the ship speed, $v(s)$: the one from the derivative of the range finder position estimation ($v(s)+n_1(s)$) and the one from the integration of IMU acceleration ($v(s)+n_2(s)$). Each of these signals has their own spectral characteristics, here modeled by their different noise levels, $n_1(s)$ and $n_2(s)$. If both signals have complementary spectral characteristics, transfer functions may be chosen in such a way as to minimize speed estimation error. The general requirement is that one of the transfer functions complement the other. Thus, for both measurements of the speed signal [7]:

$$H_1(s) + H_2(s) = 1.$$

This will allow the signal component to pass through the system undistorted since the output of the system will always sum to one. In this case, n_1 is predominantly high-frequency noise and n_1 is low-frequency noise; these two

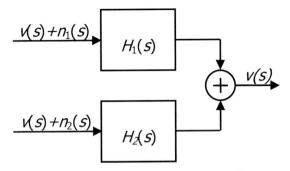

Figure 3.5   Speed control. complementary filter.

noise sources have complementary spectral characteristics. Choosing $H_1(s)$ to be a low-pass filter, and $H_2(s)$ a high-pass filter, both with the same, suitably cut frequency, the output v will not suffer from any delay in dynamic response due to low-pass filtering, and will be free of both high- and low-frequency noise.

3.2.4 Track-Keeping Control

Regarding heading control, IMU and sonar data are used for keeping the model centered and in course along the towing tank. In case these values are not accurate enough, heading control may be switched to a manual mode and an external RC transmitter could be used for course keeping. At first, the signals of the sonars to maintain the model centered on the tank and a Kalman filter taken data from the IMU were used to keep the course, the magnetometers' signals being of primary importance in this Kalman filter.

During testing, this arrangement showed to be not very effective because the steel rails of the carriage, placed all along at both sides of the tank, induced a shift in the magnetometer signals when the ship model was not perfectly centered at the tank. In addition, the magnetometers were also sensible to the electrical power lines coming across the tank. For these reasons only the sonar signals were used to keep both course and position, with the help of the relative position to the carriage, which have been used to keep the speed constant anyway.

3.2.5 Other Components

The power for all the elements is provided by two 12 V D.C batteries, placed abaft in the ship, providing room in their locations for longitudinal and transverse mass adjustment. These batteries have enough capacity for a whole day of operation.

The main propulsion system block, consisting of the already mentioned brushless motor (1), electronic speed control (ESC, not in view), two-stage planetary gearbox (2), rotational speed and torque sensor (3), elastic couplings (4), and an output shaft with Kardan couplings (5), is represented in Figure 3.6. The disposition of these elements in a single block allows a modular implementation of the whole system and simplifies the operations needed to install and uninstall it from a given ship model.

Finally, two adjustable weights have been placed in both sides of the ship model, and another one has been placed forward in the centerline. In both

Figure 3.6 Propulsion system block.

cases, enough room has been left as to allow transversal and vertical mass adjustment. Moreover, two sliders with 0.5 kg weights have been installed for fine tuning of the mass distribution.

3.3 Testing

As mentioned above, the proposed model is mainly intended to be used in seakeeping tests in towing tanks, where the carriage may interfere in the motion of the ship. The application of the proposed model to one of these tests, aimed at predicting and preventing the appearance of parametric roll resonance, will be presented in this section.

As it has been already described, parametric roll resonance could generate large roll motions and lead to fatal consequences. The need for a detection system, and even for a guidance system, has been recursively stated by the maritime sector [8]. However, the main goal is to obtain systems that can, in a first stage, detect the appearance of parametric roll resonance, but that, in a second stage, can prevent it from developing.

As mentioned, there are some specific conditions that have to be present for parametric roll to appear, regarding both ship and waves characteristics. Wave encounter frequency should be around twice the ship natural roll frequency, their amplitude should be over a certain threshold that depends on ship characteristics, and their wavelength should be near the ship´s length. Ship roll damping should be small enough not to dissipate all the energy that is generated by the parametric excitation. And finally, ship restoring arm variations due to wave passing along the hull, should be large enough as to counteract the effect of roll damping.

If parametric roll resonance wants to be prevented, it is obvious that there are three main approaches that could be used, and that consist in avoiding the presence of at least one of the aforementioned conditions.

The first alternative consists in acting on the ship damping components, or using stabilizing forces that oppose the heeling arms generated during the resonance process. The second alternative, aimed at reducing the amplitude of restoring arm variations, necessarily implies introducing modifications in hull forms. Finally, the third alternative is focused on avoiding wave encounter frequency being twice the ship natural roll frequency. If we consider a ship sailing at a given loading condition (and so with a constant natural roll frequency) in a specific seaway (frequency and amplitude), the only way of acting on the frequency ratio is by modifying encounter frequency. So, this alternative would be based on ship speed and heading variations that take the vessel out of the risk area defined by $\frac{\omega_e}{\omega_n} = 2$. This last approach is the one adopted in this work.

3.3.1 Prediction System

Regarding the prediction system, Artificial Neural Networks (ANN) have been used as a roll motion forecaster, exploiting their capabilities as nonlinear time series estimators, in order to subsequently detect any parametric roll event through the analysis of the forecasted time series [9]. In particular, multilayer perceptron neural networks (MPNNs) have been the structures selected to predict ship roll motion in different situations. The base neural network architecture used in this work is a multilayer perceptron network with two hidden layers, 30 neurons per layer, and one output layer. This initial structure is modified by increasing the number of neurons per layer, or the number of layers, in order to compare the prediction performance of different architectures and according to the complexity of the cases under analysis.

The network is fed with time series of roll motion, which are 20 seconds long and sampled at a frequency $F_s = 2$ Hz; hence the input vector $\mathbf{x}^0$ has 40 components. The network has only one output, which is the one step-ahead prediction. By substituting part of the input vector with the network output values, and recursively executing the system, longer predictions can be obtained from the initial 20 seconds. However, as the number of iterations grows, the prediction performance deteriorates accordingly.

The length of the roll time series has been selected taking into account two factors. On the one hand, the natural roll periods of the vessels chosen for the testing. In the loading conditions considered for these tests, this value is

7.48 seconds. On the other hand, parametric roll fully develops in a short period of time, usually no more than four rolling cycles [10].

The training of the artificial neural network forecaster has been achieved by using the roll time series obtained in the towing tank tests that will be described in the following section.

These algorithms have been implemented within the ship onboard control system, and their performance analyzed in some of the aforementioned towing tank tests.

3.3.2 Prevention System

Once parametric resonance has been detected, the control system should act modifying the model speed or heading in order to prevent the development of the phenomenon. In order to do so, the concept of stability diagrams has been applied. These diagrams display the areas in which, for a given loading condition and as a function of wave height and encounter frequency – natural roll frequency ratio and for different forward speeds, parametric roll takes place. From the analysis of these regions, the risk state of the ship at every moment could be determined and its speed and heading modified accordingly to take the model out of the risk area.

In order to set up these stability diagrams, a mathematical model of the ship behavior has been developed, validated, and implemented within the ship control system in order to compute the stability areas for the different loading conditions tested.

This model is a one degree of freedom nonlinear model of the ship roll motion:

$$(I_{xx} + A_{44}) \cdot \ddot{\phi} + B_{44,\,T} \cdot \dot{\phi} + C_{44}(\phi, t) = 0,$$

where I_{xx} and A_{44} are respectively the mass and added mass moments of inertia in roll, $B_{44,\,T}$ represents the nonlinear damping term and $C_{44}(\phi, t)$ is the time varying nonlinear restoring coefficient. In this case, the moment and the added moment of inertia have been obtained by measuring the natural roll frequency from a zero speed roll decay test, carried out with the developed ship scale model.

While trying to model parametric roll resonance, it has been seen that both heave and pitch motions have a clear influence in the appearance of the phenomenon, together with the effects of the wave moving along the hull. So, including the influence of these factors is of paramount importance for a good performance of the mathematical model. In this work, the influence of these factors has been taken into account while computing the restoring term, by adopting the quasi-static "look up table" approach, described by [11] and required by the ABS Guidelines [12] for modelling the variation of the ship restoring capabilities in longitudinal waves.

Moreover, and regarding the roll damping, it has been shown that it is essential for a good simulation of large amplitude roll motion, to consider this parameter as highly nonlinear. In order to account for this fact, roll damping has been decomposed into two terms, one linear component, which is supposed to be dominant at small roll angles, and a quadratic one, which is necessary to model the effects of damping at large roll angles. This approach has also been applied by many authors for modelling parametric roll, that is, [13], with accurate results. In order to determine the two damping coefficients $(B_{44,\,a}, B_{44,\,b})$ in the most accurate way, roll decay tests for different forward speeds have been carried out in still water for the loading condition under analysis. The procedure followed for determining the damping coefficients from these tests is that described in [14].

Once the model was correctly set up and validated, it was executed for different combinations of wave height and frequency, the maximum roll amplitude for these time series was computed and the stability diagrams developed.

3.3.3 Towing Tank Tests and Results

The proposed system has been used to perform different tests, some of which have been published elsewhere [15]. The main objective of these tests was to analyze, predict and prevent the phenomenon of parametric roll resonance. It is in this sort of tests, characterized by large amplitude oscillations in both roll and pitch motions, where the proposed system performs best as it can take information on board without disturbing the free motion of the ship model.

To illustrate the influence of the towing device on the measures obtained in this kind of tests, Figure 3.7 is presented. On it, the pitch and roll motions of a conventional carriage-towed model (Figure 3.8) in a similar parametric rolling test, are included.

As it can be observed, the ship pitch motion presents a series of peaks (the most relevant in seconds 140, 180, 220 and 300), which are due to the interference of the towing device. These interferences not only influence the model pitch motion, but could also affect the development of parametric roll and so, the reliability of the test.

The tests campaign has been carried out in the towing tank of the Escuela Técnica Superior de Ingenieros Navales of the Technical University of Madrid.

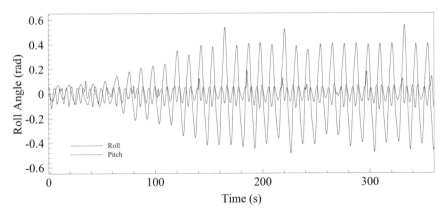

Figure 3.7 Roll and pitch motions in parametric roll resonance. Conventional carriage-towed model.

Figure 3.8 Conventional carriage-towed model during testing.

This tank is 100 meters long, 3.8 meters wide and 2.2 meters deep. It is equipped with a screen type wave generator, directed by a wave generation software, capable of generating longitudinal regular and irregular waves according to a broad set of parameters and spectra. The basin is also equipped with a towing carriage able to develop a speed of up to 4.5 m/s. As it has already been mentioned, in this test campaign, trials at different forward speeds and also at zero speed have been carried out.

Regarding the zero speed runs, in order to keep the model in position and to try to avoid as much as possible any interferences of the restraining devices in the ship motions, two fixing ropes with two springs have been tightened to the sides of the basin and to a rotary element fixed to the model bow. Moreover, another restraining rope has been fitted between the stern of the model and the towing carriage, stowed just behind of it. However, this last rope has been kept loose and partially immersed, being enough for keeping the model head to seas without producing a major influence on its motion. In the forward speed test runs, the model has been left sailing completely free, with the exception of a security rope that would be used just in case the control of the ship was lost.

In order to set the adequate speed for each test run, a previous calibration for different wave conditions has been carried out to establish the needed engine output power (engine servo command) for reaching the desired speed as a function of wave height and frequency. The exact speed value developed in each of the test runs has been measured by following the model with the towing carriage, which has provided the instantaneous speed along the run. The calibration curve has been updated as the different runs were completed, providing more information for subsequent tests. However and considering that the model is free to move in the six degrees of freedom, the instantaneous speed of the model may be affected by surge motion, especially at the conditions with highest waves.

A total number of 105 test runs have been carried out in head regular waves. Different combinations of wave height (ranging from 0.255 m to 1.245 m model scale) and ratio between encounter frequency and natural roll frequency (from 1.80 to 2.30) have been considered for three different values of forward speed (Froude numbers 0.1, 0.15 and 0.2) and zero speed, and two different values of metacentric height (0.370 m and 0.436 m). From the whole set of test runs, 55 correspond to the 0.370 m GM case, while 50 correspond to a GM of 0.436 m.

The results obtained from the different test runs have been applied for determining ship roll damping coefficients and validating the performance of the mathematical model described in the previous subsection, for validating

Figure 3.9 Proposed model during testing.

the correctness of the so computed stability diagrams, and for training and testing the ANN detection system.

In addition to this, the results of pitch and roll motion obtained with the proposed model are presented in Figure 3.10, for the sake of comparison with the results obtained with the conventional model (Figure 3.7). As it can be seen, the pitch time series doesn't present the peaks observed in the conventional

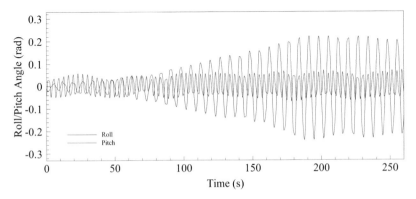

Figure 3.10 Roll and pitch motions in parametric roll resonance. Proposed model.

model measurements, as no interference between model and carriage occurs in this case.

3.3.3.1 Mathematical model validation

As a first step, the results obtained from the towing tank tests have been used for validating the performance of the roll motion nonlinear mathematical model to predict the amplitude of the resulting roll motion.

In most cases, the results obtained from the model are quite accurate, and correctly simulate the roll motion of the ship both in resonant and non-resonant conditions.

Examples for both situations could be seen in Figures 3.11 and 3.12:

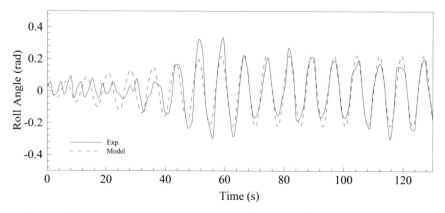

Figure 3.11   Comparison between experimental and numerical data. Resonant case.

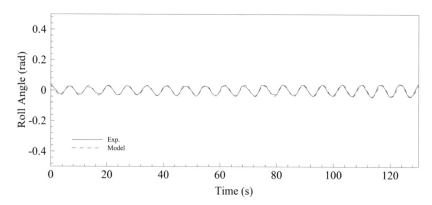

Figure 3.12   Comparison between experimental and numerical data. Non-Resonant case.

3.3.3.2 Validation of stability diagrams

Once the model has been validated, it has been recursively executed for a set of combinations of different wave frequencies and heights, covering the frequency ratio range from 1.70 to 2.40 and wave heights from 0.20 to 1.20 m, at each operational condition under consideration (four speeds), in order to set up the stability diagrams. Once computed, the results have been compared to those obtained in the towing tank tests. To illustrate this comparison, the results obtained from these experiments (Figure 3.13) have been superimposed on the corresponding mathematical plots (Figure 3.14); light dots represent non-resonant conditions, while resonance is shown by the dark dots.

As can be seen in Figure 3.14, the shape of the unstable region matches the results obtained in the towing tank tests, that together with the accurate results obtained while computing roll amplitude, show a good performance of the mathematical model.

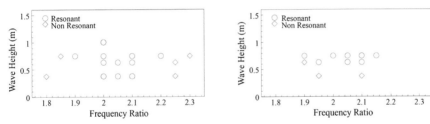

Figure 3.13 Experimental stability diagrams. Fn = 0.1. (Left), Fn = 0.15 (Right), GM = 0.370 m.

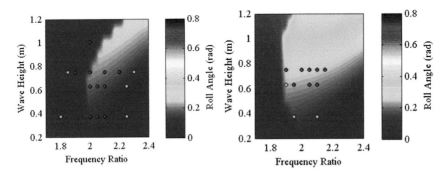

Figure 3.14 Comparison between experimental and numerical stability diagrams. Fn = 0.1. (Left), Fn = 0.15 (Right), GM = 0.370 m.

3.3.3.3 Prediction system tests

To forecast the onset and development of the parametric roll phenomena, some standard perceptron ANNs have been used. Several ANN architectures were tested and the overall best results were obtained with 3 layers of 30 neurons each.

The training cases for the ANNs have been obtained from the experiments that have been carried out with different values of wave frequency and amplitude at a Froude number of 0.2, consisting of 14 time series averaging a full scale length of 420 seconds. IMU output data was sampled by the on-board computer at a rate of 50 data sets per second. As for this particular ship model, the roll period of oscillation is of several seconds, the data used for training the ANNs was under-sampled to 2 Hz, which was more than enough to capture the events and permit reducing the size of the data set used. This resulted in a total number of 11169 training cases.

Encounter frequency – natural roll frequency ratio ranged from 1.8 to 2.3, implying that there would be cases where parametric roll was not present. Due to this fact, the performance of the system in a condition where only small roll amplitudes appear due to small transversal excitations or when roll motions decrease (cases that were not present in the mathematical model tests as no other excitation was present apart from head waves) could be evaluated.

The tests have been performed on both regular and irregular waves in cases ranging from small to heavy parametric roll. In regular waves, the RMS error when predicting 10 seconds ahead has been of the order of 10^{-3} in cases presenting large roll amplitudes and it reduces to 10^{-4} in cases with small amplitudes.

Two examples of these predictions are shown in Figures 3.15 and 3.16, which include one case with fully developed parametric roll, and another one without any resonant motions. On these figures, the ANN forecast is represented by the dotted line, while the real data is represented by full lines. Figure 3.15 is a typical example of a case presenting large amplitudes, while in Figure 3.16 no resonance motion takes place. As can be observed in both cases, the results are very good.

Further details of the characteristics and performance of the forecasting ANN system have been presented by the authors in [16]. There, the forecasting system has been implemented on a ship model instrumented with accelerometers and tested by using standard towing tank methods. The data used for the Figure 3.7 plot has been obtained during this testing campaign.

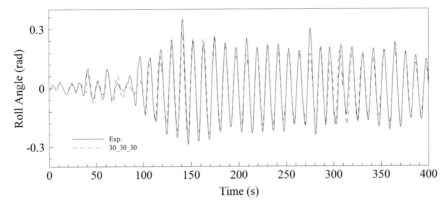

Figure 3.15   Forecast results. 30 neuron, 3 layer MP. 10 seconds prediction. Resonant case.

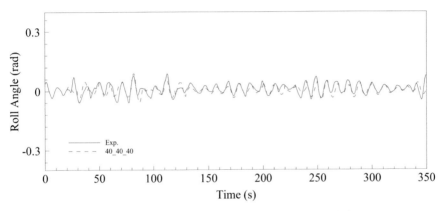

Figure 3.16   Forecast results. 30 neuron, 3 layer mp. 10 Seconds prediction. Non Resonant case.

3.4 Conclusions and Future Work

The development and implementation of an autonomous scale ship model for towing tank testing has been presented as well as some of the results obtained with it during a real towing tank test campaign. The system is aimed to be installed on board of self-propelled models, acting as an autopilot that controls speed and track, the latter by maintaining course and keeping the model centered in the tank. It also has an IMU with a 3-axis accelerometer, a gyroscope and a magnetometer and, in addition, it measures the torque, rotational speed and propulsive force at the propeller. A model ship so instrumented is able to move without any restriction along any of its six degrees of motion; consequently, the system produces optimal measurements even in tests cases presenting motions of large amplitude.

At is present development stage, the system only needs to use the towing carriage as a reference for speed and position. A more advanced version that could eliminate the use of this carriage is under development. This towing carriage, together with its rails, propulsion and instrumentation, is a very costly piece of hardware. The final version of the system could be constructed at a fraction of this cost and it will be a true towless towing tank, as it will allow performing any standard towing tank test without the need of an actual tow.

References

[1] T. I. Fossen, 'Handbook of Marine Craft Hydrodynamics and Motion Control', John Wiley & Sons, 2011.

[2] W. N. France, M. Levadou, T. W. Treakle, J. R. Paulling, R. K. Michel and C. Moore, 'An investigation of head-sea parametric rolling and its influence on container lashing systems', Marine Technology, vol. 40(1), pp. 1–19, 2003.

[3] A. Francescutto, 'An experimental investigation of parametric rolling in head waves', Journal of Offshore Mechanics and Arctic Engineering, vol. 123, pp. 65–69, 2001.

[4] I. Drummen, 'Experimental and Numerical Investigation of Nonlinear Wave-Induced Load Effects in Containerships considering Hydroelasticity', PhD Thesis, Norwegian University of Science and Technology, Norway, 2008.

[5] International Towing Tank Conference (ITTC), 'Testing and Extrapolation Methods. Loads and Responses, Stability. Model Tests on Intact Stability', ITTC 7.5–02-07–04.1. 2008.

[6] R. Brown, P. Hwang, 'Introduction to Random Signals and Applied Kalman Filtering, Second Edition', John Wiley and Sons Inc., 1992.

[7] E. R. Bachmann, 'Inertial and Magnetic Tracking of Limb Segment Orientation for Inserting Humans into Synthetic Environments', PhD Thesis, Naval Posgraduate School, USA, 2000.

[8] K. Døhlie, 'Parametric Roll - a problem solved?', DNV Container Ship Update, 1, 2006.

[9] R. M. Golden, 'Mathematical methods for neural network analysis and design', The MIT Press, 1996.

[10] International Maritime Organization (IMO), 'Revised Guidance to the Master for Avoiding Dangerous Situations in Adverse Weather and Sea Conditions (Vol. IMO MSC.1/Circ. 1228)', IMO Maritime Safety Committee, 82nd session, 2007.

[11] G. Bulian, 'Nonlinear parametric rolling in regular waves - a general procedure for the analytical approximation of the GZ curve and its use in time domain simulations', Ocean Engineering, 32 (3–4), pp. 309–330, 2005.

[12] American Bureau of Shipping (ABS), 'Guide for the Assessment of Parametric Roll Resonance in the Design of Container Carriers', 2004.

[13] M. A. S. Neves, C. A. Rodríguez, 'On unstable ship motions resulting from strong non-linear coupling', Ocean Engineering, 33 (14, 15), 1853–1883, 2006.

[14] Y. Himeno, 'Prediction of Ship Roll Damping. A State of the Art', Department of Naval Architecture and Marine Engineering, The University of Michigan College of Engineering, USA, 1981.

[15] M. Míguez González, F. López Peña, V. Díaz Casás, L. Pérez Rojas, 'Experimental Parametric Roll Resonance Characterization of a Stern Trawler in Head Seas', Proceedings of the 11th International Conference on the Stability of Ships and Ocean Vehicles, Athens, 2012.

[16] F. López Peña, M. Míguez González, V. Díaz Casás, R. J. Duro, D. Pena Agras, 'An ANN Based System for Forecasting Ship Roll Motion', Proceedings of the 2013 IEEE International Conference on Computational Intelligence and Virtual Environments for Measurement Systems and Applications, Milano, Italy, 2013.

4

Autonomous Knowledge Discovery Based on Artificial Curiosity-Driven Learning by Interaction

K. Madani, D. M. Ramik and C. Sabourin

Images, Signals & Intelligent Systems Lab. (LISSI / EA 3956) University PARIS-EST Créteil (UPEC) –Sénart-FB Institute of Technolog, Lieusaint, France
Corresponding author: K. Madani <madani@u-pec.fr>

Abstract

In this work, we investigate the development of a real-time intelligent system allowing a robot to discover its surrounding world and to learn autonomously new knowledge about it by semantically interacting with humans. The learning is performed by observation and by interaction with a human. We describe the system in a general manner, and then we apply it to autonomous learning of objects and their colors. We provide experimental results both using simulated environments and implementing the approach on a humanoid robot in a real-world environment including every-day objects. We show that our approach allows a humanoid robot to learn without negative input and from a small number of samples.

Keywords: Visual saliency, autonomous learning, intelligent system, artificial curiosity, automated interpretation, semantic robot-human interaction.

4.1 Introduction

In recent years, there has been a substantial progress in robotic systems able to robustly recognize objects in the real world using a large database of pre-collected knowledge (see [1] for a notable example). There has been, however, comparatively less advance in the autonomous acquisition of such knowledge: if contemporary robots are often fully automatic, they are rarely fully autonomous in their knowledge acquisition. If the aforementioned substantial progress is commonsensical regarding the last-decades' significant developments in methodological and algorithmic approaches relating visual information processing, pattern recognition and artificial intelligence, the languishing in the machine's autonomous knowledge acquisition is also obvious regarding the complexity of the additional necessary skills to achieve such "not algorithmic" but "cognitive" task.

Emergence of cognitive phenomena in machines have been and remain an active part of research efforts since the rise of Artificial Intelligence (AI) in the middle of the last century, but the fact that human-like machine-cognition is still beyond the reach of contemporary science only proves how difficult the problem is. In fact, nowadays there are many systems, such as sensors, computers or robotic bodies, that outperform human capacities; nonetheless, none of the existing robots can be called truly intelligent. In other words, robots sharing everyday life with humans are still far away. Somewhat, it is due to the fact that we are still far from fully understanding the human cognitive system. Partly, it is so because it is not easy to emulate human cognitive skills and complex mechanisms relating those skills. Nevertheless, the concepts of bio-inspired or human-like machine-cognition remain the foremost sources of inspiration for achieving intelligent systems (intelligent machines, intelligent robots, etc...). This is the way we have taken (e.g. through inspiration from biological and human knowledge acquisition mechanisms) to design the investigated human-like machine-cognition based system able to acquire high-level semantic knowledge from visual information (e.g. from observation). It is important to emphasize that the term "cognitive system" means here that characteristics of such a system tend to those of human cognitive systems. This means that a cognitive system, which is supposed to be able to comprehend the surrounding world on its own, but whose comprehension would be non-human, would afterward be incompetent of communicating about it with its human counterparts. In fact, human-inspired knowledge representation and human-like communication (namely semantic) about the acquired knowledge become key points expected from such a system. To achieve the aforementioned capabilities, such a cognitive system should thus be able to develop its own high-level representation of facts from low-level visual information (such as images). Accordingly to the expected autonomy, the processing from the "sensory level" (namely visual level) to the "semantic level" should be performed solely by the robot, without human supervision. However, this does not mean excluding interaction with humans, which is, on the contrary, vital for any cognitive system, be it human or machine. Thus, the investigated system has to share its perceptual high-level knowledge of the world with the human by interacting with him. The human on his turn shares with the cognitive robot his knowledge about the world using natural speech (utterances) completing observations made by the robot.

In fact, if a humanoid robot is required to learn to share the living space with its human counterparts and to reason about it in "human terms", it has to face at least two important challenges. One, coming from the world itself, is the vast number of objects and situations the robot may encounter in the real world. The other one comes from humans' richness concerning various ways they use to address those objects or situations using natural language. Moreover, the way we perceive the world and speak about it is strongly culturally dependent. This is shown in [2] regarding usage of color terms by different people around the world, or in [3] regarding cultural differences in description of spatial relations. A robot supposed to defeat those challenges cannot rely solely on a priori knowledge that has been given to it by a human expert. On the contrary, it should be able to learn on-line, within the environment in which it evolves and by interaction with the people it encounters in that environment (see [4] for a survey on human-robot interaction and learning and [5] for an overview of the problem of anchoring). This learning should be completely autonomous, but still able to benefit from interaction with humans in order to acquire their way of describing the world. This will inherently require that the robot has the ability of learning without an explicit negative evidence or "negative training set" and from a relatively small number of samples. This important capacity is observed in children learning the language [6]. This problem has been addressed to different degrees in various works. For example, in [7] a computational model of word-meaning, acquisition by interaction is presented. In [8], the authors present a computational model for the acquisition of a lexicon describing simple objects. In [9], a humanoid robot is taught to associate simple shapes to human lexicon. In [10], a humanoid robot is taught through a dialog with untrained users with the aim to learn different objects and to grasp them properly. More advanced works on robots' autonomous learning and dialog are given by [11, 12].

In this chapter, we describe an intelligent system, allowing robots (as for example humanoid robots) to learn and to interpret the world in which they evolve using appropriate terms from human language, while not making use of a priori knowledge. This is done by word-meaning anchoring based on learning by observation and by interaction with its human counterpart. Our model is closely inspired by human infants' early-ages learning behaviour (e.g.

see [13, 14]). The goal of this system is to allow a humanoid robot to anchor the heard terms to its sensory-motor experience and to flexibly shape this anchoring according to its growing knowledge about the world. The described system can play a key role in linking existing object extraction and learning techniques (e.g. SIFT matching or salient object extraction techniques) on one side, and ontologies on the other side. The former ones are closely related to perceptual reality, but are unaware of the meaning of objects they are treated, while the latter ones are able to represent complex semantic knowledge about the world, but, they are unaware of the perceptual reality of concepts, which they are handling.

The rest of this chapter is structured as follows. Section 4.2 describes the architecture of the proposed approach. In this section, we detail our approach by outlining its architecture and principles, we explain how beliefs about the world are generated and evaluated by the robot and we describe the role of human-robot interaction in the learning process. Validation of the presented system on colors learning and interpretation, using simulation facilities, is reported in Section 4.3. Section 4.4 focuses on the implementation and validation of the proposed approach on a real robot in a real-world environment. Finally, Section 4.5 discusses the achieved results and outlines future work.

4.2 Proposed System and Role of Curiosity

Curiosity is a key skill for human cognition and thus it appears as an appealing concept in conceiving artificial systems that gather knowledge, especially when they are supposed to gather knowledge autonomously. Accordingly to Berlyne's Theory of human curiosity [15], two kinds of curiosities stimulate the human's cognitive mechanism. The first one is the so-called "perceptual curiosity", which leads to increased perception of stimuli. It is a lower-level function, more related to perception of new, surprising or unusual sensory input. It relates reflexive or repetitive perceptual experiences. The other one is called "epistemic curiosity", which is more related to the "desire for knowledge that motivates individuals to learn new ideas, to eliminate information-gaps, and to solve intellectual problems.

According to [16] and [17], the general concept of the presented architecture could include one unconscious visual level which may contain a number of Unconscious Cognitive Functions (UCF) and one conscious visual level which may contain a number of Conscious Cognitive Functions (CCF). Conformably with the aforementioned concept of two kinds of curiosity, an example of knowledge extraction from visual perception, involving both kinds of curiosity, is shown on Figure 4.1. The perceptual curiosity motivates or stimulates what we call the low-level knowledge acquisition and concerns "reflexive" (unconscious) processing level. It seeks "surprising" or "attention-drawing" information in given visual data. The task of the perceptual curiosity is realized by perceptual saliency detection mechanisms. This gives the basis for operation of high-level knowledge acquisition, which is stimulated by epistemic curiosity. Being previously defined as the process that motivates to "learn new ideas, eliminate information-gaps, and solve intellectual problems": as those relating the interpretation of visual information or the belief's generation concerning the observed objects.

The problem of learning brings an inherent problem of distinguishing the pertinent sensory information and the impertinent one. The solution to this task is not obvious even if we achieve joint attention in the robot. This is illustrated on Figure 4.2. If a human points to one object (e.g. an apple) among many others, and describes it as "red", the robot still has to distinguish which of the detected colors and shades of the object the human is referring to.

To achieve correct anchoring in spite of such an uncertainty, we adopt the following strategy. The robot extracts features from important objects found in the scene along with the words the tutor used to describe the objects. Then, the robot generates its beliefs about which word could describe which feature. The beliefs are used as organisms in a genetic algorithm. Here, the appropriate

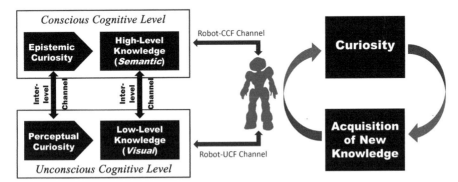

Figure 4.1 General Bloc-diagram of the proposed curiosity driven architecture (left) and principle of curiosity-based stimulation-satisfaction mechanism for knowledge acquisition (right).

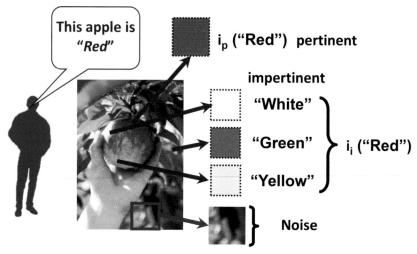

Figure 4.2   A Human would describe this Apple as "Red" in spite of the fact, that this is not the only visible color.

fitness function is of major importance. To calculate the fitness, we train a classifier based on each belief and using it, we try to interpret the objects the robot has already seen. We compare the utterances pronounced by the human tutor in the presence of each such an object with the utterances the robot would use to describe it based on the current belief. The closer the robot's description is to the one given by the human, the higher the fitness is. Once the evolution has been finished, the belief with the highest fitness is adopted by the robot and is used to interpret occurrences of new (unseen) objects. On Figure 4.3, important parts of the system proposed in this paper are depicted.

4.2.1 Interpretation from Observation

Let us suppose a robot equipped with a sensor observing the surrounding world. The world is represented as a set of features $I = \{i_1, i_2, \cdots, i_k\}$, which can be acquired by this sensor [18]. Each time the robot makes an observation o, a human tutor gives it a set of utterances U_m describing the important (e.g. salient) objects found. Let us denote the set of all utterances ever given about the world as U. The observation o is defined as an ordered pair $o = \{I_l, U_m\}$, where $I_l \subseteq I$, expressed by Equation (4.1), stands for the set of features obtained from observation and $U_m \subseteq U$ is a set of utterances (describing o) given in the context of that observation. i_p denotes the pertinent information for a given u (i.e. features that can be described

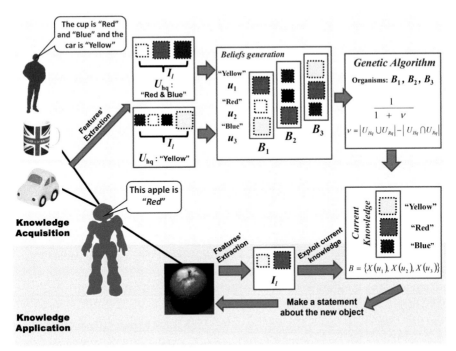

Figure 4.3 A Human would describe this Toy-frog as green in spite of the fact, that this is not the only visible color.

semantically as *u* in the language used for communication between the human and the robot), i_i the impertinent information i_i (i.e. features that are not described by the given *u*, but might be described by another $u_i \in U$) and sensor noise ε. The goal for the robot is to distinguish the pertinent information present in the observation from the impertinent one and to correctly map the utterances to appropriate perceived stimuli (features). In other words, the robot is required to establish a word-meaning relationship between the uttered words and its own perception of the world. The robot is further allowed to interact with the human in order to clarify or verify its interpretations.

$$I_l = \bigcup_{U_m} i_p(u) + \bigcup_{U_m} i_i(u) + \varepsilon. \tag{4.1}$$

Let us define an interpretation $X(u) = \{u, I_j\}$ of an utterance *u* an ordered pair where $I_j \subseteq I$ is a set of features from *I*. So, the belief *B* is defined according to Equation (4.2) as an ordered set of *X(u)* interpreting utterances *u* from *U*.

$$B = \{X(u_1), \cdots, X(u_n)\}. \tag{4.2}$$

According to the criterion expressed by (4.3), one can calculate the belief B, which interprets in the most coherent way the observations made so far: in other words, by looking for such a belief, which minimizes across all the observations $o_q \in O$ the difference between the utterances U_{Hq} made by the human, and those utterances U_{Bq}, made by the system by using the belief B. Thus, B is a mapping from the set U to I: all members of U map to one or more members of I and no two members of U map to the same member of I.

$$\arg\min_{B} \left(\sum_{q=1}^{|O|} |U_{Hq} - U_{Bq}| \right). \tag{4.3}$$

Figure 4.4 gives, through example, an alternative scheme of the defined notions and their relationship. It depicts a scenario in which two observations o_1 and o_2 are made corresponding to two description U_1 and U_2 of those observations, respectively.

On first observation, features i_1 and i_2 were obtained along with utterances u_1 and u_2, respectively. Likewise for the second observation, features i_3, i_4 and i_5 were obtained along with utterance u_3. In this example, it is easily visible that the entire set of features $I = \{i_1, \cdots, i_5\}$ contains two sub-sets I_1 and I_2. Similarly the ensemble of whole utterances $\{u_1, u_2, u_3\}$ give the

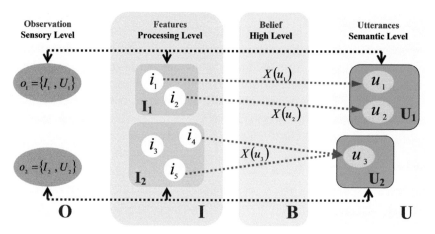

Figure 4.4   Bloc-diagram of relations between observations, features, beliefs and utterances in sense of terms defined in the text.

set U_H and their sub-sets U_1 and U_2 refer to the corresponding observations (e.g. $q \in \{1, 2\}$). In this view, an interpretation $X(u_1)$ is a relation of u_1 with a set of features from I (namely I_1). Then, a belief B is a mapping (relation) from the set U to I. All members of U map to one or more members of I and no two members of U are associated to the same member of I.

4.2.2 Search for the Most Coherent Interpretation

The system has to look for a belief B, which would make the robot describe a particular scene with utterances as close and as coherent as possible to those made by a human on the same scene. For this purpose, instead of performing the exhaustive search over all possible beliefs, we propose to search for a suboptimal belief by means of a genetic algorithm. For doing that, we assume that each organism within it has its genome constituted by a belief, which, results into genomes of equal size $|U|$ containing interpretations $X(u)$ of all utterances from U. The task of coherent belief generation is to generate beliefs which are coherent with the observed reality.

In our genetic algorithm, the genomes' generation is a belief generation process generating genomes (e.g. beliefs) as follows. For each interpretation $X(u)$ the process explores the whole the set O. For each observation $o_q \in O$, if $u \in U_{Hq}$ then features $i_q \in I_j$ (with $I_j \subseteq I$) are extracted. As described in (1), the extracted set contains pertinent as well as impertinent features. The coherent belief generation is done by deciding, which features $i_q \in I_j$ may possibly be the pertinent ones. The decision is driven by two principles. The first one is the principle of "proximity", stating that any feature i is more likely to be selected as pertinent in the context of, if its distance to other already selected features is comparatively small. The second principle is the "coherence" with all the observations in O. This means that any observation $o_q \in O$, corresponding to $u \in U_{Hq}$, has to have at least one feature i assigned into I_j of the current $X(u) = \{u, I_j\}$ [19]. Thus, it is both the similarity of features and the combination of certain utterances, describing observations from O (characterized by certain features), that guide the belief generation process. These beliefs may be seen as "informed guesses" on the interpretation of the world as perceived by the robot.

To evaluate a given organism, a classifier is trained, whose classes are the utterances from U and the training data for each class $u \in U$ are those corresponding to $X(u) = \{u, I_j\}$, i.e. the features associated with the given u in the genome. This classifier is used through the whole set O of observations, classifying utterances $u \in U$ describing each $o_q \in O$ according to its extracted features. Such a classification results in the set of utterances U_{Bq} (meaning that a belief B is tested regarding the qth observation). The fitness function evaluating the fitness of each above-mentioned organism is defined as "disparity" between U_{Bq} and U_{Hq} (defined in the previous subsection) which is computed according to the Equation (4.4), where v is given by Equation (4.5) representing the number of utterances that are not present in both sets U_{Bq} and U_{Hq}, which means that they are either missed or are superfluous utterances interpreting the given features.

$$D\left(v\right) = \frac{1}{1+v} \qquad (4.4)$$

$$v = \left| U_{Hq} \bigcup U_{Bq} \right| - \left| U_{Hq} \bigcap U_{Bq} \right|. \qquad (4.5)$$

At the end of the above-described genetic evolution process, the globally best fitting organism is chosen as the belief that best explains the observations O made (by the robot) so far about the surrounding world.

4.2.3 Human-Robot Interaction

Human beings learn both by observation and by interaction with the world and with other human beings. The former is captured in our system in the "best interpretation search" outlined in previous subsections. The latter type of learning requires that the robot be able to communicate with its environment and is facilitated by learning by observation, which may serve as its bootstrap. In our approach, the learning by interaction is carried out in two kinds of interactions: human-to-robot and robot-to-human. The human-to-robot interaction is activated anytime the robot interprets the world wrongly. When the human receives a wrong response (from the robot), he provides the robot a new observation by uttering the desired interpretation. The robot takes this new corrective knowledge about the world into account and searches for a new interpretation according to this new observation. The robot-to-human interaction may be activated when the robot attempts to interpret a particular feature. If the classifier trained with the current belief classifies the given feature with a very low confidence, then this may be a sign that this feature is a borderline example. In this case, it may be beneficial to clarify its true nature. Thus, led by epistemic curiosity, the robot asks its human counterpart to make an utterance about the uncertain observation. If the robot does not interpret according to the utterance given by the human (the robot's interpretation was wrong), this observation is recorded as new knowledge and a search for the new interpretation is started.

Using these two ways of interactive learning, the robot's interpretation of the world evolves both in amount, covering increasingly more phenomena as they are encountered, and in quality, shaping the meaning of words (utterances) to conform with the perceived world.

4.3 Validation Results by Simulation

In the simulated environment, images of real-world objects were presented to the system alongside with textual tags describing colors present on each object. The images were taken from the Columbia Object Image Library database (COIL: it contains 1000 color images of different views of 100 objects). Five fluent English speakers were asked to describe each object in terms of colors. We restricted the choice of colors to "Black", "Gray", "White", "Red", "Green", "Blue" and "Yellow", based on the color opponent process theory [20]. The tagging of the entire set of images was highly coherent across the subjects. In each run of the experiment, we have randomly chosen a tagged set. The utterances were given in the form of text extracted from the descriptions. The object was accepted as correctly interpreted if the system's and the human's interpretations were equal.

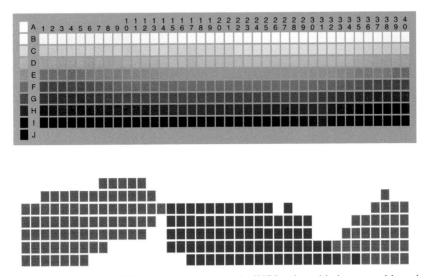

Figure 4.5 Upper: the WCS color table. lower: the WCS color table interpreted by robot taught to distinguish warm (marked by red), cool (blue) and neutral (white) colors.

The rate of correctly described objects from the test set was approximately 91% after the robot had fully learned. Figure 4.5 gives the result of interpretation by the system of the colors of the WCS table regarding "Warm" and "Cool" colors.

Figure 4.6 shows the learning rate versus the increasing number of exposures of each color. It is pertinent to emphasize the weak number of learned examples (required examples) leading to a correct recognition rate

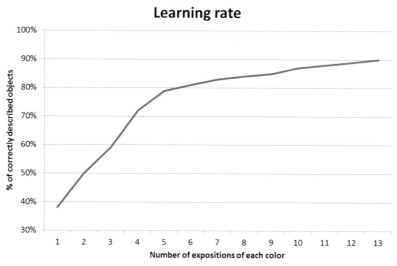

Figure 4.6 Evolution of number of correctly described objects with increasing number of exposures of each color to the simulated robot.

Figure 4.7 Examples of obtained visual colors' interpretations (lower images) and corresponding original images (upper images) for several testing objects from COIL database.

of 91%. Finally, Figure 4.7 gives an example of objects' colors interpretation by the system.

4.4 Implementation on Real Robot and Validation Results

The validation of the proposed system has been performed on the basis of both simulation of the designed system and by an implementation on a real humanoid robot[1]. As real robot we have considered the NAO robot (a small humanoid robot from Aldebaran Robotics) which provides a number of facilities such as onboard camera (vision), communication devices and onboard speech generator. The fact that the above-mentioned facilities were already available offers a huge save of time, even if those faculties remain quite basic in that kind of robot.

Although the usage of the presented system is not specifically bound to humanoid robots, it is pertinent to state two main reasons why a humanoid robot has been used for the system's validation. The first reason for this is that from the definition of the term "humanoid", a humanoid robot aspires to make its perception close to the human one, entailing a more human-like experience of the world. This is an important aspect to be considered in the context of sharing knowledge between a human and a robot. Some aspects of this problem are discussed in [21]. The second reason is that humanoid robots are specifically designed in order to interact with humans in a "natural" way by using a loudspeaker and microphone set. Thus, required facilities for bi-directional communication with humans through speech synthesis and speech recognition are already available on such kinds of robots. This is of major importance when speaking is a central item for natural human-robot interaction.

4.4.1 Implementation

The core of the implementation's architecture is split into five main units: Communication Unit (CU), Navigation Unit (NU), Low-level Knowledge Acquisition Unit (LKAU), High-level Knowledge Acquisition Unit (HLAU) and Behavior Control Unit (BCU). Figure 4.8 illustrates the bloc-diagram of the implementation's architecture. The aforementioned units control NAO robot (symbolized by its sensors, its actuators and its interfaces in Figure 4.8)

---

[1]A video capturing different parts of the experiment may be found online on: http://youtu.be/W5FD6zXihOo

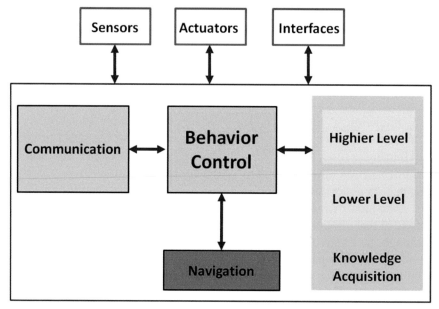

Figure 4.8 Block diagram the implementation's architecture.

through its already available hardware and software facilities. In other words, the above-mentioned architecture controls the whole robot's behavior.

The purpose of NU is to allow the robot to position itself in space with respect to objects around it and to use this knowledge to navigate within the surrounding environment. Capacities needed in this context are obstacle avoidance and determination of distance to objects. Its sub-unit handling spatial orientation receives its inputs from the camera and from the LKAU. To get to the bottom of the obstacle avoidance problem, we have adopted a technique based on ground color modeling. Inspired by the work presented in [22], color model of the ground helps the robot to distinguish free-space from obstacles. The assumption is made that obstacles repose on ground (i.e. overhanging and floating objects are not taken into account). With this assumption, the distance of obstacles can be inferred from monocular camera data. In [23], some aspects of distance estimation from a static monocular camera have been mentioned, proffering the robot the capacity to infer distances and sizes of surrounding objects.

The LKAU ensures gathering of visual knowledge, such as detection of salient objects and their learning (by the sub-unit in charge of salient object detection) and sub-recognition (see [18, 24]). Those activities are carried out mostly in an "unconscious" manner, that is, they are run as an automatism in "background" while collecting salient objects and learning them. The learned knowledge is stored in Long-term Memory for further use.

The HKAU is the center where the intellectual behavior of the robot is constructed. Receiving its features from the LKAU (visual features) and from the CU (linguistic features), this unit processes the belief generation, the most coherent beliefs emergence and constructs the high-level semantic representation of acquired visual knowledge. Unlike the LKAU, this unit represents conscious and intentional cognitive activity. In some way, it operates as a baby who learns from observation and from verbal interaction with adults about what he observes developing in this way his own representation and his own opinion about the observed world [25].

The CU is in charge of robot communications. It includes an output communication channel and an input communication channel. The output channel is composed of a Text-To-Speech engine which generates human voice through loudspeakers. It receives the text from the BCU. The input channel takes its input from a microphone and through an Automated Speech Recognition engine (available in NAO) the syntax and semantic analysis (designed and incorporated in BCU) it provides the BCU labeled chain of strings representing the heard speech. As it has been mentioned, the syntax analysis is not available on NAO. Thus it has been incorporated in BCU. To perform syntax analysis, the TreeTagger tool is used. Developed by the ICL at University of Stuttgart, the TreeTagger tool is a tool for annotating text with part-of-speech and lemma information. Figure 4.9 shows, through a simple example of an English phrase, the operational principle of syntactic analysis performed by this tool. "Part-of-speech" row gives tokens explanation and the "Lemma" row shows lemmas output, which is the neutral form of each word in the phrase. This information along with known grammatical rules for creation of English phrases may further serve to determine the nature of the phrase as

Phrase	Robots	are	our	friends
Tokens	NNS	VBS	PP$	NNS
Part-of-speech	Noun, plural	Verb, present	Possessive pron	Noun, plural
Lemma	robot	be	our	friend

Figure 4.9 Example of English phrase and the corresponding syntactic analysis output generated by treetagger.

declarative (for example: "This is a Box"), interrogative (for example: "What is the name of this object?") or imperative (for example: "Go to the office"). It can be also used to extract the subject, the verb and other parts of speech, which are further processed in order to make emerge the appropriate action by the robot. Figure 4.10 gives the flow diagram of communication between the robot and a human as it has been implemented in this work.

The BCU plays the role of a coordinator of robot's behavior. It handles data flows and issues command signals for other units, controlling the behavior of the robot and its suitable reactions to external events (including its interaction with humans). BCU received its inputs from all other units and returns its outputs to each concerned unit including robot's devices (e.g. sensors, actuators and interfaces) [25].

4.4.2 Validation Results

A total of 25 every-day objects was collected for experimental purposes of (Figure 4.11). They have been randomly divided into two sets for training and for testing. The learning set objects were placed around the robot and then a human tutor pointed to each of them calling it by its name. Using its 640x480 monocular color camera, the robot discovered and learned the objects from its surrounding environment containing objects from the above-mentioned set.

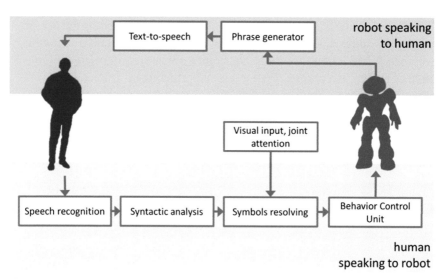

Figure 4.10   Flow diagram of communication between a robot and a human which is used in this work.

Figure 4.11 Everyday objects used in the experiments in this work.

The first validation involving the robot has aimed at verifying the leaning, color interpretation, interaction with human and description abilities of the proposed (e.g. investigated) system. To do this, the robot has been asked to learn a subset of the 25 objects: in terms of associating the name of each detected object to that object. At the same time, a second learning process has been performed involving the interaction with the tutor who has successively pointed the above-learned objects describing (e.g. telling) to the robot the color of each object. Here below, an example of the Human-Robot interactive learning is reported:

- **Human**: [pointing a red aid-kit] "This is a first-aid-kit!"
- **Robot**: "I will remember that this is a first-aid-kit."
- **Human**: "It is red and white."
- **Robot**: "OK, the first-aid-kit is red and the white."

After learning the names and colors of the observed objects, the robot is asked to describe a number of objects including also some of the already learned objects but in a different posture (for example the yellow chocolate box presented in reverse posture, etc.) and a number of still unseen objects (as for example a red apple or a white teddy-bear). The robot has successfully described, in coherent linguistics, the presented seen and unseen objects. Here below is an example of Human-Robot interaction during the recognition phase:

- **Human**: [pointing the unseen white teddy-bear]: "Describe this!"
- **Robot**: "It is white!"
- **Human**: [pointing the already seen, but reversed, yellow chocolate box]: "Describe this!"
- **Robot**: "It is yellow!"
- **Human**: [pointing the unseen apple]: "Describe this!"
- **Robot**: "It is red!"

Figure 4.12 shows two photographs of the above-reported experimental validation, where the robot completes its knowledge by interacting with a human and learning from him. Figure 4.13 shows another two photographs where the robot describes different objects it could detect within its surrounding environment.

In another experiment, taking into account the previously learned objects, among which the robot has detected and has learned a black book (the only available book in the learning set), the robot has been asked to search for the book in a new (e.g. unseen) environment where a number of different objects, including the above-mentioned black book, have been placed with two new books (unseen) as well as other objects looking similar to a book. The robot's behavior and Human-Robot interaction during this experiment is given here below:

- **Human**: "Go find the book!"
- **Robot**: "OK, I am searching for the book!"

Figure 4.12   Tutor pointing an aid-kit detected by robot describing its name and color to the robot (left-side picture). Pointing, in the same way, another visible objects detected by robot, tutor describes them to the robot (Right-Side picture).

Figure 4.13 Tutor pointing a yellow chocolate box which has been seen, interpreted and learned (by the robot) in terms of colors then asking the robot to describe the chosen object (left-side picture). Tutor pointing an unseen white teddy-bear asking the robot to describe the chosen object (right-side picture).

- **Robot**: [after visually exploring the new surrounding environment]: "I can see three books!"
- **Human**: "Get the red one!"
- **Robot**: [after turning toward the "red and grey" book and after walking and reaching the aforementioned book's location] "I am near the red book!"

Figure 4.14 shows two photographs of the robot during its searching for the required "red" book. Besides the correct recognition of the desired object (asked by the tutor), what is pertinent and interesting to note is that the robot also found two other unseen books. What is also very pertinent to emphasize, and very interesting, is that even if there is no "red" book in that environment, the robot has correctly interpreted the fact that the red book required by the human was the "red and grey" book: the only book that may coherently

Figure 4.14 Images from a video sequence showing the robot searching for the book (left-side picture) and robot's camera view and visualization of color interpretation of the searched object (right-side picture).

be considered as "red" by the human. A video showing the experimental validation may be found on http://youtu.be/W5FD6zXihOo. More details of the presented work with complementary results can be found in [19, 25].

4.5 Conclusions

This chapter has presented, discussed and validated a cognitive system for high-level knowledge acquisition from visual perception based on the notion of artificial curiosity. Driving as well the lower as the higher levels of the presented cognitive system, the emergent artificial curiosity allows such a system to learn in an autonomous manner new knowledge about the unknown surrounding world and to complete (enrich or correct) its knowledge by interacting with a human. Experimental results, performed as well on a simulation platform as using the NAO robot, show the pertinence of the investigated concepts as well as the effectiveness of the designed system. Although it is difficult to make a precise comparison due to different experimental protocols, the results we obtained show that our system is able to learn faster and from significantly fewer examples than most of more-or-less similar implementations.

Based on the results obtained, it is thus justified to say that a robot endowed with such artificial curiosity-based intelligence will necessarily include autonomous cognitive capabilities. With respect to this, the further perspectives regarding the autonomous cognitive robot presented in this chapter will focus on integration of the investigated concepts in other kinds of robots, such as mobile robots. There, it will play the role of an underlying system for machine cognition and knowledge acquisition. This knowledge will be subsequently available as the basis for tasks proper for machine intelligence such as reasoning, decision making and an overall autonomy.

References

[1] D. Meger, P. E. Forssén, K. Lai, S. Helmer, S. McCann, T. Southey, M. Baumann, J. J. Little and D. G. Lowe, 'Curious George: An attentive semantic robot', Robot. Auton. Syst., vol. 56, no. 6, pp. 503–511, 2008.
[2] P. Kay, B. Berlin and W. Merrifield, 'Biocultural Implications of Systems of Color Naming', Journal of Linguistic Anthropology, vol. 1, no. 1, pp. 12–25, 1991.

[3] M. Bowerman, 'How Do Children Avoid Constructing an Overly General Grammar in the Absence of Feedback about What is Not a Sentence?', Papers and Reports on Child Language Development, 1983.

[4] M. A. Goodrich and A. C. Schultz, 'Human-robot interaction: a survey', Found. Trends Hum.-Comput. Interact., vol. 1, no. 3, pp. 203–275, 2007.

[5] S. Coradeschi and A. Saffiotti, 'An introduction to the anchoring problem', Robotics & Autonomous Sys., vol. 43, pp. 85–96, 2003.

[6] T. Regier, 'A Model of the Human Capacity for Categorizing Spatial Relations', Cognitive Linguistics, vol. 6, no. 1, pp. 63–88, 1995.

[7] J. de Greeff, F. Delaunay and T. Belpaeme, 'Human-robot interaction in concept acquisition: a computational model', Proc. of Int. Conf. on Development and Learning, vol. 0, pp. 1–6, 2009.

[8] P. Wellens, M. Loetzsch and L. Steels, 'Flexible word meaning in embodied agents', Connection Science, vol. 20, no. 2–3, pp. 173–191, 2008.

[9] J. Saunders, C. L. Nehaniv and C. Lyon, 'Robot learning of lexical semantics from sensorimotor interaction and the unrestricted speech of human tutors', Proc. of 2nd International Symposium on New Frontiers in Human-Robot Interaction, Leicester, pp. 95–102, 2010.

[10] Lütkebohle, J. Peltason, L. Schillingmann, B. Wrede, S. Wachsmuth, C. Elbrechter and R. Haschke, 'The curious robot - structuring interactive robot learning', Proc. of the 2009 IEEE international conference on Robotics and Automation, Kobe, pp. 2154–2160, 2009.

[11] T. Araki, T. Nakamura, T. Nagai, K. Funakoshi, M. Nakano and N. Iwahashi, 'Autonomous acquisition of multimodal information for online object concept formation by robots', Proc. of IEEE/ IROS, pp. 1540–1547, 2011.

[12] D. Skocaj, M. Kristan, A. Vrecko, M. Mahnic, M. Janicek, G.-J. M. Kruijff, M. Hanheide, N. Hawes, T. Keller, M. Zillich and K. Zhou, 'A system for interactive learning in dialogue with a tutor', Proc. of IEEE/ IROS, pp. 3387–3394, 2011.

[13] C. Yu, 'The emergence of links between lexical acquisition and object categorization: a computational study', Connection Science, vol. 17, 3–4, pp. 381–397, 2005.

[14] S. R. Waxman and S. A. Gelman, 'Early word-learning entails reference, not merely associations', Trends in cognitive science, 2009.

[15] D. E. Berlyne, 'A theory of human curiosity', British Journal of Psychology, vol. 45, no. 3, August, pp. 180–191, 1954.

[16] K. Madani, C. Sabourin, 'Multi-level cognitive machine-learning based concept for human-like artificial walking: Application to autonomous stroll of humanoid robots', Neurocomputing, S.I. on Linking of phenomenological data and cognition, pp. 1213–1228, 2011.

[17] K. Madani, D. Ramik, C. Sabourin, 'Multi-level cognitive machine-learning based concept for Artificial Awareness: application to humanoid robot's awareness using visual saliency', J. of Applied Computational Intelligence and Soft Computing,. DOI: 10.1155/2012/354785, 2012. (available on: http://dx.doi.org/10.1155/2012/354785).

[18] D. M. Ramik, C. Sabourin, K. Madani, 'A Machine Learning based Intelligent Vision System for Autonomous Object Detection and Recognition', J. of Applied Intelligence, Springer, Vol. 40, Issue 2, pp. 358–374, 2014.

[19] D-M. Ramik, C. Sabourin. K. Madani, 'From Visual Patterns to Semantic Description: a Cognitive Approach Using Artificial Curiosity as the Foundation', Pattern Rgognition Letters, Elsevier, vol. 34, no. 14, pp. 1577–1588, 2013.

[20] M. Schindler and J. W. v. Goethe, 'Goethe's theory of colour applied by Maria Schindler', New Knowledge Books, East Grinstead, Eng., 1964.

[21] V. Klingspor, J. Demiris, M. Kaiser, 'Human-Robot-Communication and Machine Learning', Applied Artificial Intelligence, pp. 719–746, 1997.

[22] J. Hofmann, M. Jngel, M. Ltzsch, 'A vision based system for goal-directed obstacle avoidance used in the rc'03 obstacle avoidance challenge', Lecture Notes in Artificial Intelligence, Proc. of 8th International Workshop on RoboCup, pp. 418–425, 2004.

[23] D. M. Ramik, C. Sabourin, K. Madani, 'On human inspired semantic slam's feasibility', Proc. of the 6th International Workshop on Artificial Neural Networks and Intelligent Information Processing (ANNIIP 2010), ICINCO 2010, INSTICC Press, Funchal, pp. 99–108, 2010.

[24] R. Moreno, D. M. Ramik, M. Graña, K. Madani, 'Image Segmentation on the Spherical Coordinate Representation of the RGB Color Space', IET Image Processing, vol. 6, no. 9, pp. 1275–1283, 2012.

[25] D. M. Ramik, C. Sabourin, K. Madani, 'Autonomous Knowledge Acquisition based on Artificial Curiosity: Application to Mobile Robots in Indoor Environment', J. of Robotics and Autonomous Systems, Elsevier, Vol. 61, no. 12, pp. 1680–1695, 2013.

5

Information Technology for Interactive Robot Task Training Through Demonstration of Movement[1]

F. Kulakov and S. Chernakova

Laboratory of Informational Technologies for Control and Robots,
St. Petersburg Institute for Informatics and Automation of the Russian
Academy of Sciences, Russia
Corresponding author: F. Kulakov <kufelix@yandex.ru>

Abstract

Remote robot control (telecontrol) includes the solution of the following routine problems: surveillance of the remote working area, remote operation of the robot situated in the remote working area, as well as pre-training of the robot. The current paper describes a new technique for robot control using intelligent multimodal human-machine interfaces (HMI). The first part of the paper explains the robot control algorithms, including testing of the results of learning and of movement reproduction by the robot. The application of the new training technology is very promising for space robots as well as for modern assembly plants, including the use of micro-and nano-robots.

Keywords: Robot, telecontrol, task training by demonstration, human-machine interfaces.

5.1 Introduction

The concept of telesensor programming (TSP) and relevant task-oriented robot control techniques for use in space robotics was first proposed by G. Herzinger [1].

---

[1]The paper is published with financial support from the Russian Foundation for Basic Research, projects 14-08-01225-a, 15-07-04415-a, 15-01-02021-a.

Within the framework of the ROTEX program, implemented in April 1993 for the SPACE-LAB space station, a simulation environment for multisensory semiautonomous robot systems, with powerful man-machine interfaces (laser range finders, 3D-stereo graphics and force/torque effort reflection), was developed. This allowed the space robot manipulator to be remotely programmed (teleprogrammed) from Earth.

The solution for the problem of remote control under non-deterministic delays in the communications channel is based on the use of TSP with training by demonstration for the sensitized robot.

Tasks such as assembly, joining of connectors and catching flying objects were practiced. Actually, it was the first time that a human remotely trained a robot through direct movement demonstration using a graphic model with robot sensor simulation.

The effectiveness of interactive control (demonstration training) is highlighted in all cases of the application of pre-training technology to space and medical robots, as the most natural way to transfer the operator's experience (SKILL TRANSFER) in order to ensure autonomous robot manipulator operation in a complex non-deterministic environment [2–5].

However, in these studies it was only possible to conduct training with the immediate recording of the movement trajectory positioning data and the possibility of motion correction as per the signals from the robot's sensors.

These studies did not solve the problem of complex robot motion representation as a certain data structure that is easily adjustable by humans, or "independently" modified by the autonomous robot, depending on changes in the remote environment.

The current paper describes a new information technology-based approach for interactive training by demonstration of the human operator's natural hand movements based on motion representation in the form of a frame-structured model (FSM).

Here, a frame means a description of the shape of motion with indications of its metric characteristics, methods and sequence of execution of the separate parts of the movement. Training by demonstration means intelligent robot manipulator programming aimed at training the robot for autonomous work with the objects (among the objects) without point-to-point trajectory recording. That is, by providing only separate fragments of movement in the training stage, and sequentially executing them, depending on the task.

In order to train a robot manipulator to move among objects it was suggested to use a remotely operated camera, fixed to the so-called "sensitized glove". This allows not only the registration of position and orientation of the hand in space, but also the position of the object (experimental models) models' characteristic points relative to the camera on the hand.

5.2 Conception and Principles of Motion Modeling

5.2.1 Generalized Model of Motion

A variety of robot motion types in the external environment (EE), including manipulation of items (objects and tools) as well as the complexity and variability of EE configurations, are typical for aerospace, medical, industrial, technological and assembly operations.

Let us consider the problem of training the robot manipulator to perform motion relative to EE objects in two cases: examination motion and manipulative motion. The main issue in forming the motion patterns, set, in this case, by the motions of the operator's head and arm, is to have a method for recording and reproducing the three-dimensional trajectories of the robot manipulator grip relative to EE objects.

The problem of the alignment of the topology and the semantics of objects, well known in geographic information systems (GIS), is basically close to the problem of motion modeling and route planning in robotics.

In the case of navigational routing tasks using intelligent GIS, the authors basically consider motion along a plane (on the surface of the sphere) or several planes (echelon gratings). Moreover, in most cases, the moving object is taken as a mathematical point, not having its own orientation in space.

The motion path configuration in space often does not matter, so routing is carried out over the shortest distance. Thus, while following the curvature of the relief, the motion tracks its shape.

For object shape modeling and motion formation, we propose using a common structured description language, which considers that the object shape model is defined and described by a frame of its elements, and the motion trajectory model is described by a frame of descriptions of the elementary motions. It is important to note that the elementary motions (fragments) can be given appropriate names and be considered to be the language operators, providing the possibility of describing robot actions in a rather compact manner.

For interactive motion demonstration robot training, we propose using a combination of the EE (MEE) objects' shape models and the motion shape models (MFM). In this case, the generalized frame-structured model (FSM) is defined as a method for storing information not only about the shape of the EE objects, but also about the shape of the motion trajectory.

The description language used in FSM is a multilevel hierarchical system of frames, similar to M. Minski frames [6], containing a description of the shape elements, metric characteristics and methods and procedures for working with these objects. MFM, as one of the FSM components, stores the structure of the shape of motion trajectories demonstrated by the human operator during the process of training the robot to perform specified movements [7, 8].

The generalized FSM of the remotely operated robot IE includes:

- Object models, models of the objects' topology (location) in a particular IE (MIE);
- Models of different typical motions and topology models (interrelations, locations) of these movements in a particular IE (MIE).

It is also proposed to store, in the MIE, the coordinates and images of objects from positions convenient both for remote-camera observation (which enables the most accurate measurement of the coordinates of the characteristic features of object images) and for grabbing objects with the robot gripper (Figure 5.1) [9].

Training of motion can be regarded as a transfer of knowledge of motor, sensory, and behavioral skills from a human operator to the robot control system (RCS), which in this case should be a multimodal man-machine interface (MMI), developed to the greatest possible extent (intelligent) to provide adequate and effective perception of human actions. Consequently,

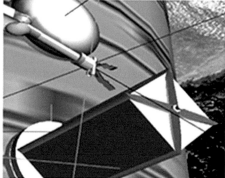

Figure 5.1   Images of the Space Station for two positions: "Convenient for observation" and "Convenient for grabbing" objects with the virtual manipulator.

it is assumed that a generalized model of description of the robot knowledge on the EE based on the FSM will be created, including the robot itself and its possible (necessary) actions within it.

The preliminary results of the research on algorithms and technologies for the robot manipulator task training by demonstration, using the motion description in the form of MFM, are presented below.

5.2.2 Algorithm for Robot Task Training by Demonstration

In order to task-train the robot by demonstration, a special device, the so-called "sensitized glove," is put on the hand of the trainer. It is equipped with a television camera and check points (markers) [10].

This allows the execution of two functions simultaneously (Figure 5.2):

- Using the television camera on the glove, record the image and determine the coordinates of the objects' characteristic points, over which the hand of the human operator moves;
- Using the sensors of the intelligent MMI system, determine the spatial position and orientation of the hand in the work location by means of 3–4 check points (markers) on the glove.

Considering the processes for task-training a robot to perform elementary operations and reproducing these operations, an important feature is revealed. This feature consists in the fact that algorithms for training and reproduction present fragments, which are used in different operations without

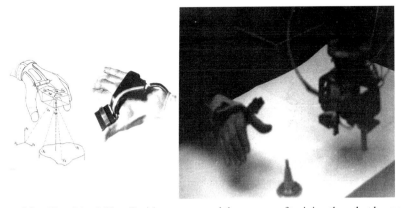

Figure 5.2 "Sensitized Glove" with a camera and the process of training the robot by means of demonstration.

modifications or with very minor changes, and may also be repeated several times in a single operation.

From among the various movements of the robot manipulator, most of them can be represented as a sequence of a limited number of elementary motions (motion fragments), for example:

- Transfer motion of the gripper along an arbitrary complex trajectory $g = g(l)$ from the current position to a certain final position;
- Correction motion, using the sequence of characteristic points (CP) of the EE objects' images, as input information;
- Surveillance movement in the process by which the following are sequentially created: matrices of the gripper position T_b, T_{b1}, T_{b2}, joint coordinate vectors g_b, g_{b1}, g_{b2}, and geometric trajectory $g = g(l)$;
- Movement to a convenient position for surveillance;
- Movement to a convenient position for grabbing;
- Movement for "tracking" the object (approaching the object);
- Movement to grab the object.

In traditional training systems using one or the other method, a sequence of points of the motion trajectory of the robot gripper is obtained. It can be represented as a function of some parameter l, which can be considered as the preliminary result of training the robot to perform the fragment of the gripper movement from one position to the other:

$$g = g\left(l\right), l_b \leq l \leq l_e$$

$$g_b = g\left(l_b\right), g_e = g\left(l_e\right),$$

where: l_b – parameter of the trajectory in the initial position, l – parameter of the trajectory in the current position, l_e– parameter of the trajectory in the final position.

In this case, the training algorithm for performing motions ensures the formation of geometric trajectory $g(l)$ and includes the following:

- Formation of a sequence of triplets of the two-dimensional vectors $x_{imb}^{(1)}$, $x_{imb}^{(2)}$, $x_{imb}^{(3)}$; $x_{imI}^{(1)}$, $x_{imI}^{(2)}$, $x_{imI}^{(3)}$; $\ldots$; $x_{ime}^{(1)}$, $x_{ime}^{(2)}$, $x_{ime}^{(3)}$, conforming to the image positions of the 3 CP on the object during training;
- Formation of the sequence T_b, T_I, T_{II}, $\ldots$, T_e of the matrices of the glove position;
- Solution of systems of equations (5.1):

$$x_{im}^{(i)} = (x_{im1}^{(i)}, x_{im2}^{(i)}) = k^{(i)}\widehat{T}X^{(i)}, \tag{5.1}$$

where: $k^{(i)}$ is a variable scale, defined as: $k^{(i)} = f/d^{(i)}\text{-}f$, where $d^{(i)}$ is the distance from the point to the TV camera showing the plane; f is the focal distance of the lens, $\widehat{T}$ is a (2x4) matrix made up of the first two rows of matrix:

$$T = \left| \begin{array}{c|c} \alpha & Xn \\ \hline 0 & 1 \end{array} \right|,$$ characterizing the rotation and displacement of the system of

coordinates (CS), in conjunction with the camera on the glove, relative to the object CS, where a is the direction cosine matrix of the reference CS rotation angle, X_n the displacement vector of the beginning of the CS and $X^{(i)}$ – the two-dimensional vectors of the position of the image of the characteristic points of the object in the image plane.

This data is sufficient to construct a sequence of matrices of the gripper positions T_b, T_1, T_{II}, ..., T_e during movement. The orientation blocks in these matrices are matrices α_b, α_1, α_{II}, ..., α_e. The block of the gripper pole position corresponds to the initial position of the gripper. According to this sequence, the geometric, and, in line with it, the temporal motion trajectory of the gripper can be built.

When teaching this action, the operator must move his hand with the glove on it in the manner in which the gripper should move during the process of the surveillance motion, whereas the position of the operator's wrist can be arbitrary and convenient for the operator.

Furthermore, for each case of teaching a new motion, it is necessary to memorize a new volume of motion information in the form of several sets of coordinates mentioned above.

When teaching the motions, e.g. IE surveillance, it is necessary to store a considerable amount of information in the memory of the robot control system (RCS), including:

- Values of matrix T, which characterize the position and orientation of the glove in the coordinate system of the operator's workstation, corresponding to initial T_b, final T_e and several intermediate T_1, T_{II}, ... gripper positions, which it must take when performing movements;
- Several images of the object from the glove-mounted TV camera, corresponding to the various gripper positions, to control the accuracy of training;
- Characteristic identification signs, characteristic points (CP) of the different images of the object, at different glove positions during the training process;
- Coordinates of the CP of the images of the object in the base coordinate system;

- Parameters of gripper opening and the permissible compressive force applied to the subject.

To reduce the amount of information and to present the motion trajectory in a language is close to the natural one, it is suggested to use a frame-structured description in the motion shape model (MFM), the basic principles of which are described in the previous papers by the authors [11, 12].

5.2.3 Algorithm for Motion Reproduction after Task Training by Demonstration

The specific feature of the robot's motion reproduction in a real IE is that fragments of elementary movements, stored during task training can follow a different sequence depending on the external conditions when reproduced. Due to the aforementioned features, it appears to be reasonable to teach the robot to do different fragments of motion in various combinations of the given fragments.

The number of applied elementary motions (fragments) increases along with the number of reproduced operations. However, this increase will be much smaller than the increase in the number of operations for which the robot is used. It is important to note that proper names can be assigned to the given elementary motions and they can be considered to be operators of the language with the help of which the robot's actions can be described in a sufficiently compact manner.

On the basis of the frame-structured description of the MFM, obtained during task training, the so-called "tuning of the MFM" for a specific task is performed before starting the reproduction of motion by the robot in a particular IE situation.

Practically, this is done by masking or selection of only those descriptions of motion in the MFD that satisfy the conditions of the task and the external conditions of the situation in the IE according to their purpose and shapes (semantic and topological features). The selected movements are automatically converted into a sequence of elementary movements $g = g(l)$.

In the case of the reproduction of the elementary motion along the trained trajectory $g = g(l)$ in systems without sensor offsetting it is necessary to construct a parameter change function $l(t)$ in the area $l_b \leq l \leq l_e$. Typically, the initial and final velocities $l(t)$ are known, and they are most commonly equal to zero:

$$l'_b = l'_e = 0.$$

In the simplest case of the formation of $l(t)$, three intervals can be singled out in it: the "acceleration" interval from the initial velocity (l'_b) to some permissible speed (l'_d), the interval of motion at a predetermined speed and the deceleration from the predetermined velocity to zero (l'_e).

During acceleration and deceleration a constant acceleration (l''_d) must take place. Its value should be such that the value of the velocity g' and acceleration g'' vectors can be physically implementable under the existing restrictions of the control vector (U) of the robot manipulator's motors.

The values of these limitations can be determined based on the consideration of the dynamic model (R) of the robot manipulator, which connects the control vector (U) to the motion dynamics vectors (g, g', g''):

$$U = R\left(g, g', g''\right).$$

In the case of the motion reproduction transfer of function $l = l(t)$, it is defined by the following ratio:

- During the acceleration interval $(0 < t = t_1)$, where $t_1 = sign\ (l'_d)\ \frac{|l'_d|}{|l''_d|}$;
- During the interval of motion at a constant velocity - $(t_1 \leq t \leq t_2)$, where
  $t_2 = t_1 + \frac{|l_e - l_b|}{|l'_d|} - \frac{|l''_d|t_1^2}{|l'_d|}$: $l(t) = l'_d(t - t_1) + l_b \frac{sign(l'_d) \cdot t_1^2}{2|l''_d|}$;
- During the deceleration interval $(t_2 \leq t \leq t_3)$, where
  $t_3 = 2t_1 + \frac{|l_e - l_b|}{|l'_d|} - \frac{|l''_d|t_1^2}{|l'_d|}$:

$$l(t - t_1) = l'_d(t - t_1) + l_b + \frac{sign(l'_d)t_1^2}{2} - \frac{sign(l'_d)l''_d(t - t_2)}{2}.$$

The reproduction of movement over time by the robot is carried out as per the implementation of the obtained function $l(t)$ in the motion trajectory $g(l)$:

$$g = g(l(t)).$$

To determine the drives' control vector $U = R(g, g', g'')$ the substitution of values g, g', g'' by the values of function $g = g(l(t))$ is carried out. This results in the formation of the control function of the motors of the robot manipulator over time.

It should be noted that a man performs natural motions with constant acceleration, in contrast to the robot manipulator, whose motions are characterized by a constant rate (speed). Therefore, the robot has to perform its motions as per its own dynamics, which differ from the dynamic properties of the human operator.

5.2.4 Verification of Results for the Task of Training the Telecontrolled (Remote Controlled) Robot

Remotely operated robots must be sufficiently autonomous and trainable to be able to efficiently perform operations in remote environments distant from the human operator. Naturally, task training for space robots must be performed in advance right here on Earth, and medical robots shall be trained out of the operation theaters.

At the same time, a possibility for the remote correction of training outcomes must be provided, for possible additional training by the human operator, located at a considerable distance from the robot in space or from a remotely controlled medical robot.

For greater reproduction reliability, it is necessary to implement an automated process control over motion reproduction by the remotely controlled robot using copies of the MFM and MEE from the RCS. Remote control over the robot movements by a human operator must be carried out using prediction, taking into consideration the potential interference and time delays in the communications channel.

Actual remote control of the space robot or the remotely operated medical robot must be carried out as follows:

- With some time advance (prediction), simultaneously with working motion execution by the robot, control over the current robot motion is performed on the simulator, which includes the MEE, MFM and the intelligent MMI system;
- Data from the RCS, arriving with delay, is reflected on the MEE and MFM and is compared to the predicted movement in order to provide the possibility of emergency correction;
- The trajectory of motion relative to the real location of the MEEs is automatically adjusted by the RCS as per sensor signals;
- By human command (or automatically by the RCS) correction of parameters and operational replacement of the motion fragments are carried out in accordance to the pre-trained alternative variants of the working motion.

After the execution by the robot of the regular working movement, actual motion trajectories in the form of a description in the language of MFM, compiled after an automatic motion analysis in the RCS, are transferred from the RCS to the human operator in the modeling environment. This information, together with the results of the real EE scanning by the robot during the robot's execution of the working movement, is to be used for correction of the MFM and MEE.

In the absence of significant corrections in the process of executing the working movement, the training is considered to be correct, understanding between the human operator and the RCS is considered to be adequate, and the results of the robot task training can be used in the future.

5.2.5 Major Advantages of Task Training by Demonstration

The proposed algorithm for task training by demonstration of the motion has a number of advantages over conventional methods of programming trajectories or motion copying, when the operator, for example, moves in space the manipulator's gripper along the desired trajectory with continuous recording of the current coordinates in the memory of the robot. Let us list the main ones.

Using the professional skills and intuitive experience of the human being. The human being, using his professional skills and intuitive experiences, demonstrates motions by hand, which are automatically analyzed by the MMI (for configuration acceptance and safety) and are conveyed to the robot in the form of a generalized MFM description. Conventional means of supervisory control, in which remote control or a joystick, are used to set the generalized command, are further developed in this case.

Simplicity and operational efficiency of training. Training is performed by simple movements of the human hand without any programming of the complex spatial displacements. It is more natural and easier for the human being to control the position of his hand during the execution of movement, than doing the same using the buttons, mouse or joystick. Experiments have shown that practically everyone can learn to control a robot through hand motion and it can be done in just a few hours. Time and cost of personnel training, for the control and training of robots are significantly reduced.

Relaxation in the requirements for motion accuracy. Instead of the exact copying and recording of arrays of coordinates of the motion trajectory during robot manipulator training, the operator gives only assignment (name) and shape of the spatial motion trajectory, including the manipulation of items, tools and EE objects. The free movement is set by the human being and is reproduced by the robot at a certainly safe distance from the objects; therefore, minute accuracy of such movements is not required. In the case where the robot gripper approaches the object, the motion is automatically adjusted according to information from the sensors, including the force-torque sensing. There is no need to copy the exact motions by the remotely operated robots, which are commonly used in partially nondeterministic EE, when there is no precise information about the location of the robot and obstacles.

Reliability of control over the autonomous robot. One of the advantages lies in the fact that there is no need for the operator to be close to the working area or to be present in the working area of the remotely operated robot, for example, inside the space station, or on the outer surface of the orbital station for operational intervention in the robot's actions, avoiding therefore delays and interferences in the communications channels. Based on the descriptions of the MFM and MEE, the intelligent RCS can automatically adjust the shape, and even the sequence of the trained motions of the robot.

Ease of control, correction and transfer of motion experience. The visual appearance of motion presentation in a MFM, its proximity to natural language of the frame structured movement description, allow reliable checking, in-flow change of composition, sequence and shape of the complex working movements directly according to the motion description text using the graphical model of robot manipulator, as well as a human model ("avatar") [13].

5.3 Algorithms and Models for Teaching Movements

5.3.1 Task Training by Demonstration of Movement among the Objects of the Environment

Robot task training and remote control is performed using the modeling environment, which contains an EE model (MEE), a model of the shape of motion (MFM) and an intelligent system for the multimodal interface (IMI), creating the so-called human "presence effect" in a remote EE using the three-dimensional virtual models and images of real items. Using the IMI, the operator can observe 3-D images on either side, like in holographs, can touch or move the virtual objects, feeling with his hand the tactile and force impact through simulations of object weight or its weight in zero-gravity environment [14, 16].

Instead of real EE models the virtual MEE image can be used, as well as a computer hand model, controlled by the hand position tracking system (HPTS), included in the IMI [17].

The process of training the robot to move among OE objects implies that the operator's hand, dressed in a sensitized glove, executes the motion in OE space that must be subsequently performed by the manipulator gripper. In order to do this, it is necessary to perform the following operations (in on-line mode) in the training stage:

- Demonstrate a fragment of the operator's hand motion among objects of the OE model or of the virtual hand model in the graphical model of the OE (MOE);
- Register through the IMM system and store in memory the fragment of motion containing the timestamps and a corresponding vector *(X)* of 6 dimensions ($x = x(l)$, $y = y(l)$, $z = z(l)$) and orientation ($\varphi_x = \varphi_x(l)$, $\varphi_y = \varphi_y(l)$, $\varphi_z = \varphi_z(l)$) of the operator's hand;
- Recognize the motion shape through the IMI system and record the results in the form of a frame-based description in MFM;
- Record the images of objects, obtained through the TV camera on the glove in the process of moving the hand and carry out recognition, identification and measurement of the coordinates of the characteristic points of the objects' images and enter this data into the MOE;
- Add to the MFM the information about the location of MOE objects relative to the glove at the moment of execution of the fragment of movement.

The position of the sensitized glove relative to objects of the EO model is determined during the process of training by demonstration by solving the so-called "navigation task". This research offers a unique solution of the navigation task for the given case [18].

While training the robot by means of demonstration, the objects (models) of the OE come in view of the TV camera fixed on the sensitized glove. There can be objects of manipulation or foreign objects (obstacles).

For the industrial and aerospace application of robots, the objects generally have regular geometric shapes, angles and edges, which may be used as characteristic features and characteristic points.

Characteristic points (CP) can be small-sized (point) details of objects, which can be easily distinguished on the image, as well as special marks or pointed sources of light. These points are the easiest way to determine the position and orientation of the camera on the sensitized glove relative to the OE objects, that is, to solve the so-called "navigation problem" during the process of robot task training.

Let us consider a case, where the position vectors of the object's CP $X^{(i)}$, ($i = 1, 2, 3$ – No. of CP) in a coordinate system associated to the object (CS) are known beforehand. Images of 3 CPs of the object ($X_{im}^{(1)}$, $X_{im}^{(2)}$, $X_{im}^{(3)}$) on the TV camera's image surface are projections of the real points

(CP1 ... CP3) on this plane in a variable scale $k^{(i)} = f / d^{(i)} - f$, inversely proportional to the distance $d^{(1)}$ from the point to the imaging plane of the lens, where f is the focal length of the lens.

Let us assume that the CS, associated with the camera lens, and, therefore, with the glove, is located as shown in Figure 5.3, i.e. axes x_1 and x_2 of the CS are located in the image plane, x_3 is perpendicular to them and is directed away from the lens towards the object. In Figure 5.3: x_1, x_2, x_3 are the axes of the coordinate system associated with the object; $x^{(1)}$, $x^{(2)}$, $x^{(3)}$ are the vectors defining the position of characteristic points in the coordinate system of the camera lens; $x_{im}^{(2)}{}_1$, $x_{im}^{(2)}{}_2$ are 2 projections of the vector from the center of the CCD matrix to the image of point 2 (this can also be shown for points 1 and 3).

Then, distance $d^{(i)}$ is equal to the projection of the i-th CP on the third axis of the CS associated with the camera: $d^{(i)} = x_{im3}^{(i)}$, and the location of the object $X^{(i)}$ in the image plane will be represented by two-dimensional vectors:

$$x_{im}^{(i)} = (x_{im1}^{(i)}, x_{im2}^{(i)}) = k^{(i)} \widehat{T} X^{(i)}, \qquad (5.2)$$

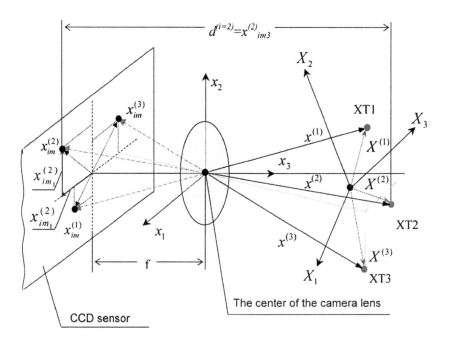

Figure 5.3   Formation of images of 3 characteristic points of the object.

where: $\widehat{T}$ is a (2×4) matrix made up of the first two rows of matrix $T = \begin{vmatrix} \alpha & Xn \\ 0 & 1 \end{vmatrix}$ characterizing the rotation and displacement of the CS associated to a camera on the glove, relative to the CS of the object, where a is a direction cosine matrix relative to the turning angle of the SC and $X_n = (X_{n1}, X_{n2}, X_{n3})$ is the displacement vector of the SC's origin,

$$X^{(i)} = (x_1^{(i)}, x_2^{(i)}, x_3^{(i)}, 1).$$

Then: $d^{(i)} = x_{im3}^{(i)} = T_3 . x^{(i)}$, where T_3 is the third row of matrix T.

It is obvious that matrix T completely determines the spatial position of the glove in the CS associated to the object, and its elements can be found as the result of solving the abovementioned navigation problem of determining the spatial position of the glove during training.

During the CP image processing, vectors $x_{im}^{(i)}$, $i = 1, 2, 3$ are determined, so the left side of Equations (5.2) is known, and these equations represent a system of six equations concerning 12 unknown elements of matrix T, which are the three components of vector X_n and nine elements of matrix a.

Since the elements of matrix a are linked by six more equations of orthogonality and orthonormality, there are a total of 12 equations, that is, as many as the unknowns. These are obviously sufficient to determine the desired matrix T.

During the "training" motion of the operator's hand at a given frequency, a procedure involving an operation for the selection of the object's CP image and an operation for calculating the values of two-dimensional vectors $x_{im}^{(i)}$, $i = 1, 2, 3$ and their position in the image plane must be performed.

As a result of these actions, a sequence of values of the vector triplets from the starting one $X_{imb}^{(i = 1,2,3)}$ to the finishing one $x_{ime}^{(i = 1,2,3)}$: $(x_{imb}^{(1)}, X_{imb}^{(2)}, X_{imb}^{(3)}); (x_{imI}^{(1)}, X_{imI}^{(2)}, X_{imI}^{(3)}); (x_{imII}^{(1)}, X_{imII}^{(2)}, X_{imII}^{(3)}); \ldots (x_{ime}^{(1)}, X_{ime}^{(2)}, X_{ime}^{(3)})$ is accumulated in the IMI database, corresponding to the sequence of the glove's positions during the movement of the operator's hand, which will be later reproduced by the robot. Each element of this sequence carries enough information to solve the navigation task, that is, to obtain the sequence $T_b, T_I, T_{II}, \ldots, T_e$ of the matrix values, which is the result of training.

After training, the robot reproduces a gripper motion based on sequence $T_b, T_I, T_{II}, \ldots, T_e$ using a motion correction algorithm based on the signals from the camera, located in the gripper, by solving the so-called "correction task" of the gripper relative to real OE objects.

5.3.2 Basic Algorithms for Robot Task Training by Demonstration

The most typical example of the robot's interaction with OE objects is the manipulation of arbitrarily oriented objects. In practice, the task of grabbing objects has several cases. The simplest case is when there is one known object. The robot must be trained to perform an operation of grabbing this object irrespective of any minor changes in its position and orientation.

A more complicated case is when the position and orientation of a known object are not known beforehand. The most typical case is when there are several known objects with a priori unknown positions and orientations. And an even more complex task is when among the known objects there are unknown objects and obstacles that may hinder the grabbing procedure.

5.3.3 Training Algorithm for the Environmental Survey Motion

During the training to perform the environmental survey motion, the operator's hand executes one or more types of search movements: rotation of the hand at two angles, zigzag motion, etc. Information about the typical search motions is recorded in the MFM. Survey motion may consist of several fragments of different movements. The sequence and shape of these motions, dependent on the task, are determined during the training phase and stored in the MFM. After the execution of separate motion fragments, a break can be taken for further analysis of the OE objects' images.

Any OE object recognition algorithm suitable for a particular purpose can be used, including a structural recognition algorithm [19].

It is necessary to note that object image analysis must include the following:

- Recognition of the target subject through a set of characteristic features $(XT1, XT2, \ldots XTk)$ that are sufficient to identify it using the description stored in the MOE;
- Selection of a set of reference points $(XT1, XT2, \ldots XTn)$ that are sufficient for navigating the robot gripper in the space of the real OE, from among a set of points $(XT1, XT2, \ldots XTk)$, usually no more than $n = 4\text{--}6$, depending on their location.

If the number of CPs observed on the object is insufficient $(k < n)$, it is necessary to perform the following search motions:

- Change the position or orientation of the camera on the glove so that $k = n$ or change the CP filter parameters in the recognition algorithm;

- Change the observation conditions or camera parameters for reliable detection of CP, such as lighting of the operator workstation or focus of the camera lens:
    - Add artificial (contrasting, color) marks on the graphical model or on the object model to recommend the use of these marks on real objects;
    - If $k \geq n$, it is possible to skip to the calculation of the spatial position and orientation of the glove relative to the object in accordance to the algorithm of the "navigation task" (see above).

Once the specified object is detected and identified and the position and orientation of the hand (glove) relative to this object is determined, the training to execute the first survey motion is deemed finished.

The purpose of the next motion the robot is trained to execute involves a gripper motion to the so-called "convenient for observation" position. In this position, the maximum identification reliability and measurement accuracy of the gripper position relative to the object are achieved.

The variants of the shift from the starting point of object detection to the position which is "convenient for observation" must be shown by the movement of the operator's wrist using his intuitive experience.

There is also an option of task training by demonstration, for survey movement, performed through natural head movements. In this case, the camera with a reference device is fixed on the operator's head.

The training process ends automatically, for example, upon a signal from the IMI system after reaching the specified recognition reliability and measurement accuracy parameters for the position of a hand or a head relative to the OE object. A training halt signal can be given by voice (using the speech recognition system of the IMI) or by pressing the button on the glove. In this case, the object coordinates defined by its image are recorded in the MFM as a vector of coordinates (X_0).

5.3.4 Training Algorithm for Grabbing a Single Object

In this case, the grabbing process consists of three movements:

- Gripper motion from the initial position to the "convenient for grabbing" the object position;
- Object gripping motion, for example, a simple translational motion along the gripper axis;
- Object grabbing motion, such as a simple closing of the gripper, exerting a given compression force.

Let's consider the task training to perform only the first action, where the operator freely moves his hand, with sensitized glove on it, to the object (model) from the initial position at the most convenient for grabbing side and sets it at a short distance from the object with the desired orientation. Information about the motion path and the hand position relative to the object, at least at the end point of the motion, is memorized in the MFM through the IMI system, which is necessary for adjusting the robot's gripper position relative to the object during the motion reproduction. It's also desirable that at least 1 or 2 CPs of the object's image get into the camera's view in the "convenient for grabbing" gripper position, so that the grabbing process can be supervised.

The training of the grabbing motion is performed along the easiest path, in order to reduce the guidance inaccuracy. If the gripper is equipped with object detection sensors, then the movement ends upon receiving signals from these sensors.

During the training to grab objects, it is necessary to memorize the transverse dimensions of the object at the spot of grabbing and the gripper compression force, sufficient for holding the object without damaging it. This operation can be implemented using additional circuit-torque and tactile sensors in the robot gripper.

In case of the presence of multiple OE objects, the training implies a more complex process of identification of the objects' images and the necessity to train additional motions, such as obstacle avoidance, changing of altitude convenient for survey in case of any shading, flashing and interference to image recognition by the camera on the glove, as well as for the camera in the robot manipulator's gripper, during the reproduction of movements.

5.3.5 Special Features of the Algorithm for Reproduction of Movements

As a result of performing the required number of training movements by the human hand, "motion experience" is formed, which is accumulated in the form of a frame-structured description in the MFM, stored in the memory of the intelligent IMI system.

The transfer of the "motion experience" from the human to the robot occurs, for example, by simply copying the MFM and MEE from the IMI memory to the RCS memory or even by transferring this data to the remotely controlled robot over communications channels. Of course, preliminary checking of training outcomes is performed, for example, on a graphical model of the robot.

Prior to the robot performing the trained movements, in accordance to the assigned task and the EE conditions, the MFM is tuned, as already mentioned (in Part I of the current paper), for example, by masking (searching) among the total volume of the MFM data for the required types of motions. Descriptions of motions, selected according to the intended purpose and shape of the trajectory, are converted by the RCS into typical motion trajectories for their reproduction by the robot in real EE.

When the robot-manipulator reproduces motions in a real EE, after training by demonstration, it is possible to execute, for example, the following typical motion trajectories:

- Survey movement in combination with EE image analysis in order to identify the object to be taken;
- Shifting of the gripper into the "convenient for observation" position;
- Corrective gripper movement relative to the object based on the signals from the robot's sensors;
- Shifting of the gripper to the "convenient for taking" position;
- Motion for grabbing the real object;
- Motion for taking the object.

Before the work starts, a complete check of the TSHP operation and telecontrol system is carried out. Then, operation of the TSHP is checked using a graphical model (GM) at the control station without using a robot and exchanging information over the communications lines. Checks of the training outcomes are performed using the surveillance MFM or manipulation MFM, located in the RCS, without switching on the robot at this moment. If necessary, additional adjustment of the MFM is performed through task training by demonstration of the natural human-hand movements and their storage in the MFM of the RCS.

The robot is switched on and it executes motions in terms of the original EE inspection, selection of objects, position selection, convenient for grabbing or convenient for visual control over object grabbing and manipulations, as well as safe obstacle avoidance, before a transition to remote control mode is performed.

The human operator sits in front of the monitor screen, which displays a graphical model or a real object image, and controls the robot through natural movements of the head and hand with the glove.

5.3.6 Some Results of Experimental Studies

The effectiveness of the proposed training technology using demonstrations of the movements, the algorithms and theoretical calculations was tested on the basis of the "virtual reality" environment at the laboratory of Information Technology in Robotics, SPIIRAS (St. Petersburg) [20].

The hardware-software environment includes:

- Two six-stage robotic manipulators of the «Puma» class, equipped with remotely operated stereo cameras and force-torque sensing;
- Models of the fragment of the space station surface, two graphic stations to work with three-dimensional models of the external environment (MEE);
- Intelligent multimodal interface (IMI) with a system for tracking hand movements (TSHP) and a system for tracking the head motions (THM) of the human operator.

The "Virtual reality" environment enables the performance of experimental studies of various information technology approaches for remote control and task training of robots:

- "Immersion technologies" of the human operator in the remote environment using the robot-like device that moves surveillance stereo cameras in the room with models of the fragment of the space station surface and containers for scientific equipment;
- "Virtual observer" technologies using the model of the freely flying machine (equipped with the surveillance camera, which allows the examination of the three-dimensional graphical model of the space station), as well as the simulation of an astronaut's work in outer space;
- Technologies for training and remotely controlling a space (medical) robot manipulator with a force-torque sensing system, which provides operational safety during manipulative operations, reflection of forces and torques on the control handle, including when working with virtual objects.

Experimental studies were performed on some algorithms for training by demonstration and remote control of a robot manipulator, including:

- Training of the robot manipulator to execute survey motions through motions of the human head;
- Scanning the surroundings by the robot and remotely operated camera on the glove;
- Using the IMI for training by demonstration of hand movements and human voice commands;
- Training the robot manipulator to grab items by demonstration of the operator's hand movements.

The motion reproduction of the robot manipulator among the real EE objects based on the use of the virtual graphical models of the EE and the robot manipulator with force-torque sensing system was also practiced in the experimental environment.

5.3.7 Overview of the Environment for Task Training by Demonstration of the Movements of the Human Head

A functional diagram of the equipment for remotely monitoring the EE is shown on Figure 5.4.

The operator, located in the control room, sets coordinates and orientations of the manipulator gripper and remotely operated camera on it using the tracking system for head position (THM). He observes the obtained EE image on the monitor screen.

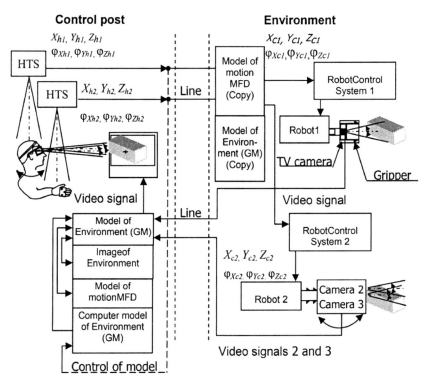

Figure 5.4 Functional diagram of robot task training regarding survey motions and object grabbing motions using THM and RCS.

Before starting, the human operator must be able to verify the THM in an off-line mode. For this purpose, a graphical model (GM) and a special communications module for controlling 6 coordinates were developed. Training the robot manipulator to execute EE surveillance motions by demonstration is carried out in the following way (Figure 5.5).

The human operator performs the EE inspection based on his personal experience in object surveillance. The robot repeats the action, using the surveillance procedure and shape of the trajectory of the human head movement. In this case, the cursor can first be moved around the obtained panoramic image, increasing (decreasing) the scale of separate fragments, and then, after accuracy validation, the actual motion of the robot manipulator can be executed.

5.3.8 Training the Robot to Grab Objects by Demonstration of Operator Hand Movements

There are several variations of the implementation of the "sensitized glove" (Figure 5.6): a remote-operated camera in the bracelet with control points and laser pointers, bracelet with active control points (infrared diodes), manipulation object - stick with the active control points [21].

When training by demonstration of human hand movements, through a sensitized glove with camera and control points, a greater range and closeness to natural movements is achieved, as compared to the use of joysticks or handle like "Master-Arm» (Figure 5.7).

This provides for the natural coordination of movements of the hand and head of the human operator. Using the head, the human controls the movement of the remotely operated surveillance camera, fixed, for example,

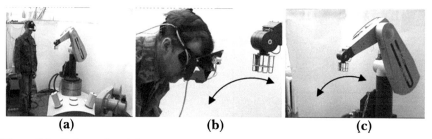

| (a) | (b) | (c) |

Figure 5.5   Robot task training to execute survey movements, based on the movements of the operator's head: Training the robot to execute survey motions to insect surroundings (a); Training process (b); Reproduction of earlier trained movements (c).

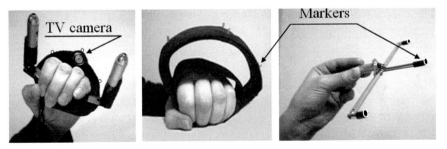

Figure 5.6 Variations of the "Sensitized Glove" construction.

Figure 5.7 Using the special glove for training the robot manipulator.

on an additional manipulator, and with the hand he controls the position and orientation of the main robot gripper (Figure 5.8).

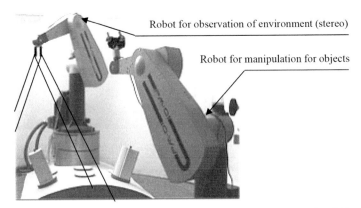

Figure 5.8 Stand for teaching robots to execute motions of surveillance and grabbing objects.

The coordination of simultaneous control using the operator's hand and head during training and remote control through natural human operator movements was put into practice in order to control the complex objects (Figure 5.9).

A new prototype of the intelligent IMI equipment with recognition of the operator's hand without markers, while performing manual control and training by demonstration of natural hand movements, was experimentally studied (Figure 5.10). In the future it is planned to continue research on new algorithms for training and remote robot control of intelligent mechatronic systems based on the use of advanced intelligent multimodal human-machine interface systems and new motion modeling principles using frame-structured

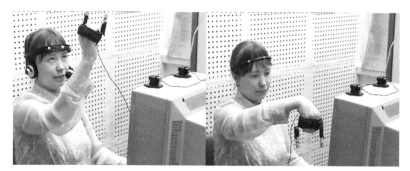

Figure 5.9   Training of motion coordination of two robot manipulators by natural movements of human head and hand.

Figure 5.10   Training with the use of a system for the recognition of hand movements and gestures without "Sensitized Gloves" against the real background of the operator's work station.

MFM descriptions, including for medical robots, mechatronic systems and telemedicine [22].

5.4 Conclusions

A new information technology approach for training robots (mechatronic systems) by demonstration of movement is based on the use of a frame-structured data representation in the models of the shape of the movements that makes it easy to adjust the movement's semantics and topology both for the human operator and for the autonomous sensitized robot.

Algorithms for training by demonstration of natural movements of the human operator's hand using a television camera, fixed on the so-called "sensitized glove", allow not only the application during the training process of graphical models of objects in surroundings but also of full-scale models, which enables the operator the possibility to practice optimal motions of the remote-controlled robots under real conditions.

It is sufficient to demonstrate the shape of a human hand movement to the intelligent system of the IMI and to enter it into the MFM, and then this movement can be executed automatically, for example, by a robot manipulator with adjustment and navigation among the surrounding objects based on the signals from the sensors.

References

[1] G. Herzinger, G. Grunwald, B. Brunner and J Heindl, 'A sensor-based telerobot system for the space robot experiment ROTEX', Proc. 2nd Internat. Symp. on Experimental Robots (ISER). Toulouse. France. 1991. June 25–27.

[2] G. Herzinger, J. Heindl, K. Landzettel and B. Brunner, 'Multisensory shared autonomy – a key issue in the space robot technology experiment ROTEX', Proc. IEEE Conf. on Intelligent Robots and Systems (IROS). Raleigh. 1992. July 7–10.

[3] Klaus H. Strobl, Wolfgang Sepp, Eric Wahl, Tim Bodenmüller, Michael Suppa, Javier F. Seara, Gerd Hirzinger, 'The DLR Multisensory Hand-Guided Device: the Laser Stripe Profiler', ICRA 2004: 1927–1932.

[4] Michael Pardowitz, Steffen Knoop, Rüdiger Dillmann, R. D. Zollner, 'Incremental Learning of Tasks From User Demonstrations, Past

Experiences, and Vocal Comments', IEEE Transactions on Systems, Man, and Cybernetics, Part B 37(2): 322–332 (2007).

[5] R. Dillmann, O. Rogalla, M. Ehrenmann, R. Zoellner and M. Bordegoni, 'Learning robot behavior and skills based on human demonstration and advice', 'The machine learning paradigm', 9th International Symposium of Robots Research (ISSR), pp. 229–238, Oct., 1999.

[6] M. Minsky, 'Frames for knowledge representation', M.: "Energia", 1979.

[7] S. E. Chernakova, F. M. Kulakov, A. I. Nechayev, 'Simulation of the outdoor environment for training by demonstration', Collected works of the SPII RAS. Issue No. 1.: St. Petersburg: SPIIRAS, 2001.

[8] S. E. Chernakova, F. M. Kulakov, A. I. Nechayev, A. I. Burdygin, 'Multiphase method and algorithm for measuring the spatial coordinates of objects for training the assembly robots', Collected works of SPIIRAS. Issue No. 1.: St. Petersburg: SPIIRAS, 2001.

[9] F. M. Kulakov, 'Technology for the Creation of Virtual Objects in the Real World', Workshop Conference, Binghamton University, NY, 4–7 March, 2002.

[10] Advanced System for Learning and Optimal Control of Assembly Robots, edited by F. M. Kulakov. SPIIRAS, St. Petersburg, 1999, pp. 76

[11] F. M. Kulakov, V. B. Naumov, 'Application of fuzzy logic and sensitized glove for programming robots', Collected works of the III International Conference "Current Problems of Informatics-98". Voronezh, pp. 59–61.

[12] C. E. Chernakova, F. M. Kulakov, A. I. Nechayev, 'Training robot by method of demonstration with the use of "sensitized" glove', Works of the First International Conference on Mechatronics and Robots. St. Petersburg, May 29 – June 2, 2000, pp. 155–164.

[13] Chernakova S. E., Timofeev A. V., Nechaev A. I., Litvinov M. V., Gulenko I. E., Andreev V. A., 'Multimodal Man-Machine Interface and Virtual Reality for Assistive Medical Systems', International Journal "Information Theories and Applications" (iTECH-2006). Varna, Bulgaria, 2006.

[14] F. M. Kulakov, S. E. Chernakova, 'Information technology of task training robots by demonstration of motions', Part I "Concept, principles of modeling movements", "Mechatronics, Automation, Control". Moscow, No. 6, 2008.

[15] F. M. Kulakov, 'Potential techniques of management supple movement of robots and their virtual models Part I', "Mechatronics, Automation, Control". Moscow. No. 11. 2003.

[16] F. M. Kulakov, 'Potential management techniques supple movement of robots and their virtual models Part II', "Mechatronics, Automation, Control". No. 1. Moscow, 2004. pp. 15–21.

[17] S. Chernakova, A. Nechaev, A. Karpov, A. Ronzhin, 'Multimodal system for hands-free PC control', 13th European signal Processing Conference EUSIPCO-2005. Turkey. 2005.

[18] F. M. Kulakov, A. I. Nechaev, S. E. Chernakova and oth., 'Eye Tracking and Head-Mounted Display/Tracking Computer Systems for the Remote Control of Robots and Manipulators', Project # 1992P, Task 5 with EOARD. St. Petersburg, 2002.

[19] A. I. Nechaev, J. S. Vorobjev, I. N. Corobkov, M. S. Olkov, V. N. Javnov, 'Structural Methods of recognition for real time systems', International Conference on Modeling Problems in Bionics "BIOMOD-92". St.-Petersburg, 1992.

[20] F. M. Kulakov, 'Technology of immersion of the virtual object into real world', Supplement to magazine "Informatsionnyie Tekhnologii", No. 10, 2004, pp. 1–32.

[21] C. E. Chernakova, F. M. Kulakov, A. I. Nechayev, 'Hardware and software means of the HMI for remote operated robots with the application of systems for tracking human operator's movements', Works of the III International Conf. "Cybernetics and technologies of XXI century", Voronezh, 2002.

[22] C. E. Chernakova, A. I. Nechayev, V. P. Nazaruk, 'Method of recording and visualization of three-dimensional X-ray images in real-time module for tasks of nondestructive testing and medical diagnostics', Magazine "Informatsionnyie Tekhnologii", No. 11, 2005, pp. 28–37.

6

A Multi-Agent Reinforcement Learning Approach for the Efficient Control of Mobile Robots

U. Dziomin[1], A. Kabysh[1], R. Stetter[2] and V. Golovko[1]

[1] Brest State Technical University, Belarus
[2] University of Ravensburg-Weingarten, Germany
Corresponding author: R. Stetter <stetter@hs-weingarten.de>

Abstract

This paper presents a multi-agent control architecture for the efficient control of a multi-wheeled mobile platform. Such control architecture is based on the decomposition of a platform into a holonic, homogenous, multi-agent system. The multi-agent system incorporates multiple Q-learning agents, which permits them to effectively control every wheel relative to other wheels. The learning process was divided into two steps: *module positioning* – where the agents learn to minimize the error of orientation, and *cooperative movement* – where the agents learn to adjust the desired velocity in order to conform to the desired position in formation. From this decomposition, every module agent will have two control policies for forward and angular velocity, respectively. Experiments were carried out with a simulation model and the real robot. Our results indicate a successful application of the purposed control architecture both in the simulation and in real robot.

Keywords: control architecture, holonic homogenous multi-agent system, reinforcement learning, Q-Learning, efficient robot control.

6.1 Introduction

An efficient robot control is an important task for the applications of a mobile robot in production. The important control tasks are power consumption optimization and optimal trajectory planning. Control subsystems should

Advances in Intelligent Robotics and Collaborative Automation, 123–146.

provide energy consumption optimization in a robot control system. Four levels of robot power consumption optimization can be distinguished:

- *Motor power consumption optimization.* Those approaches based on energy-efficient technologies of motor development that produce substantial electricity saving and improve the life of the motor drive components [1, 2];
- *Efficient robot motion.* Commonly, this is a task of an inverse kinematics calculation. But the dynamic model is usually far more complex than the kinematic model [3]. Therefore, intellectual algorithms are relevant for the optimization of a robot motion [4];
- *Efficient path planning.* Such algorithms build a trajectory and divide it into different parts, which are reproduced by circles and straight lines. The robot control subsystem should provide movement along the trajectory parts. For example, Y. Mei and others show how to create an efficient trajectory using knowledge of the energy consumption of robot motions [5]. S. Ogunniyi and M. S. Tsoeu continue this work using reinforcement learning for path search [6];
- *Efficient robot exploration.* When a robot performs path planning between its current position and its next target in an uncertain environment, the goal is to reduce repeated coverage [7].

The transportation of cargo is an actual task in modern production. Multi-wheeled mobile platforms are increasingly being used in autonomous transportation of heavy components. One of these platforms is *a production mobile robot*, which was developed and assembled at the University of Ravensburg-Weingarten, Germany [3]. The robot is illustrated in Figure 6.1(a). The platform dimensions are 1200cm in length and 800cm in width. The maximum manufacturer's payload is 500kg, battery capacity is 52Ah, and all modules drive independently.

The platform is based on four vehicle steering modules [3]. The steering module (Figure 6.1(b)) consists of two wheels powered by separate motors and behaves like a differential drive.

In this paper, we explore the problems of *formation control* and *efficient motion control* of multiple autonomous vehicle modules in circular trajectory motion. The goal is to achieve a circular motion of a mobile platform around a virtual reference beacon with optimal forward and angular speeds.

One solution to this problem, [8–10] is to calculate the kinematics of a one-wheeled robot for circle driving and then generalize it for multi-vehicle systems. This approach has shown promising modeling results.

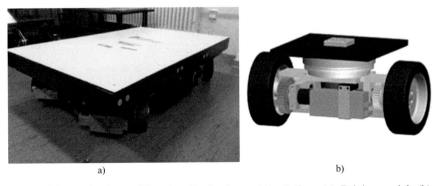

Figure 6.1 Production mobile robot: Production mobile platform (a); Driving module (b).

The disadvantage of this technique is its low flexibility and high computational complexity.

An alternative approach is to use the machine learning theory to obtain an optimal control policy. The problem of multi-agent control in robotics is usually considered as a problem of formation control, trajectory planning, distributed control and others. In this paper we use techniques from multi-agent systems theory and reinforcement learning to create the desired control policy.

The content of this paper is the following: Section 6.2 gives a short introduction to the theory of holonic, homogenous, multi-agent systems and reinforcement learning. Section 6.3 describes the steering of a mobile platform in detail. Section 6.4 describes the multi-agent decomposition of a mobile platform. Using this decomposition, we propose a multi-agent control architecture based on the model described in Section 6.2. Section 6.5 contains a detailed description of the multi-agent control architecture. The Conclusion highlights important aspects of the presented work.

6.2 Holonic Homogenous Multi-Agent Systems

A *multi-agent system* (MAS) consists of a collection of individual agents, where each agent displays a certain amount of autonomy about its actions and perception of domain, and communicates via message-passing with another agent [11, 12]. Agents act in organized structures which encapsulate the complexity of subsystems and therefore modularize its functionality. Organizations are social structures with means of conflict resolution through coordination mechanisms [13]. The overall *emergent behavior* of a multi-agent system is composed of a combination of individual agent behaviors determined by autonomous computation within each agent, and by communication among agents [14]. The field of MAS is a part of distributed AI, where each agent has a distinct problem solver for a specific task [12, 14].

6.2.1 Holonic, Multi-Agent Systems

An agent (or MAS) that appears as a single entity to the outside world but is in fact composed of many sub-agents with the same inherent structure is called *holon*, and such sub-agents are called *holonic agents* [11, 14]. The transformation of a single entity into a set of interacting subagents is called *holonic decomposition*. Holonic decomposition is an isomorphic transformation. Gerber et al. [15] show that an environment containing multiple holonic agents can be isomorphically mapped as an environment in which exactly one agent is represented explicitly, and vice versa.

For the purposes of this paper and without the loss of generality, we use terms *holon* and *holonic multi-agent system* (Holonic MAS) interchangeably, meaning that a MAS contains *exactly one* holon. In the general case, a holonic, multi-agent system (called *holarhy*) is a self-organized, hierarchical structure composed of holons [14].

A holon is always represented as a single entity to the outside world. From the perspective of the environment, a holon behaves as an autonomous agent. Only a closer inspection reveals that a holon is constructed from a set of cooperating agents. It is possible to communicate with a holon simply by sending messages to them from the environment. The most challenging problem in this design is the distribution of individual and overall computation of the holonic MAS [15].

Although it is possible to organize holonic structures in a completely decentralized manner, it is more efficient to use an individual agent to represent a holon. Representatives are called the *head of the holon*; the other agents in the holon are called the *body* [11]. In some cases, one of the already existing agents is selected as the representative of the holon. In other cases, a new agent is explicitly introduced to represent the holon during its lifetime.

The head agent represents the shared intentions of the holon and negotiates these intentions with the agents in the holon's environment, as well as with the internal agents of the holon. Only the head agent communicates with the entities outside of the holon.

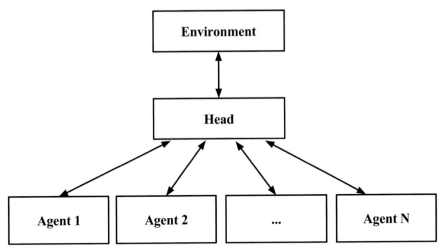

Figure 6.2 Organizational structure of a Holonic Multi-Agent System. Lines indicate the communication channels.

The organizational structure of a holonic, multi-agent system is depicted in Figure 6.2.

When agents join the holon, they surrender some of their autonomy to the head agent. The binding force that keeps the head and body in a holon together can be called a *commitments* [16]. It should be explicitly noted that agents are not directly controlled by the head agent. The agents remain autonomous entities within the holon, but they align their individual behavior with the goals of holon.

6.2.2 Homogenous, Multi-Agent Systems

For the purposes of this paper, we will consider the case when *all body* agents are *homogenous*. In a general, multi-agent scenario with homogeneous agents, there are several different agents with an identical structure (sensors, effectors, domain knowledge, and decision functions) [17]. The only differences among agents are their sensory inputs and the actual actions they take, as they are situated differently in the world [18]. Having different effector outputs is a necessary condition for MAS; if the agents all act together as a unit, then they are essentially a single agent. In order to realize this difference in output, homogeneous agents must have different sensor input as well. Otherwise, they will act identically.

Thus, the formal definition of *holonic, homogenous, multi-agent system* (H^{2}MAS) is a tuple $\mathcal{H} = \langle \mathcal{B}, h, \mathcal{C} \rangle$:

- $\mathcal{B} = \{M_1, M_2, \ldots, M_n\}$– is the set of homogenous *body* agents. Each agent is described by a tuple $M_i = \langle s, \alpha, \pi_i \rangle$, where:
- s – is the set of possible agent states, where $s_i \in s$ is the *i*-th agent current state;
- α– is the set of possible agent actions, where $a_i \in \alpha$ current action of the *i*-th agent;
- $\pi : s \rightarrow \alpha$ is the behavior policy (decision function) which maps it's state to actions;
- h – is the *head agent* representing the holon to the environment and responsible for coordinating the actions inside the holon:
- $\mathcal{S} = s^{\times n} = \{(s_1, s_2, \ldots, s_n) \mid s_i \in s \text{ for all } 1 \leq i \leq n\}$ – is a *joint state* of the holon;
- $\mathcal{A} = \alpha^{\times n} = \{(a_1, a_2, \ldots, a_n) \mid a_i \in \alpha \text{ for all } 1 \leq i \leq n\}$ is a *joint action* of the holon;
- $\pi : \mathcal{S} \rightarrow \mathcal{A}$ – global;
- $\mathcal{C}$ – is the *commitment* that defines the agreement to be inside the holon.

The learning of multi-agent systems composed of homogenous agents has a few important properties which affect the usage of such systems.

6.2.3 Approach to Commitment and Coordination in H^2 MAS

The holon is realised exclusively through cooperation among the constituent agents. The head agent is required to co-ordinate the work of the body agents to achieve the desired global behavior of H^{2}MAS by combining individual behaviors, resolving collisions, etc. In this way, a head agent serves as a *coordination strategy* among agents. The head is aware of the goals of the holon, and has access to important environmental information which allows it to act as a central point of coordination for body agents.

Since a body agent has some degree of autonomy, it may perform an unexpected action, which can lead to uncoordinated behavior within the Holon. The head agent can observe the states and actions of all subordinate agents and can fix undesired behavior using simple coordination rule: if the current behavior of the holon M_i is inconsistent with the head agent's vision, then it sends a correction message to M_i. This action by the head is known as an influence on the body. When the *body* M_i succumbs to the influence, this is called making a *commitment* to the Holon.

6.2.4 Learning to Coordinate Through Interaction

The basic idea of the selected approach for coordination is to use influences between the head and the body to determine the sequence of correct actions to coordinate behavior within the holon. The core design question is how to determine such influences in terms of received messages and how received messages affect changes of individual policies.

To answer this question we postulate that interacting agents should constantly learn optimal coordination from scratch. To achieve this, we can use *influence-based, multi-agent reinforcement learning* [18–20]. In this approach, agents learn to coordinate using reinforcement learning by exchanging rewards with each other.

In reinforcement learning, the ith agent executes an action a_i at the current state s_i. It then goes to the next state s_i' and receives a numerical reward r as feedback for the recent action [21], where $s_i, s_i' \in s, a_i \in \alpha, r \in R$. Ideally, agents should explore state space (interact with environment) to build an optimal policy π^*.

Let $Q(s, a)$ – represent a *Q-function* that reflects the quality of the specified action a in state s. Optimal policy can be expressed in terms of *optimal Q-function Q^**:

$$\pi^*(s) = \arg \max_{a \in \alpha(s)} Q^*(s, a). \qquad (6.1)$$

The initial values of Q-funcions are unknown and equal to zero. The learning goal is to approximate the *Q-function*, (e.g. to find *true Q*-values for each action in every state using received sequences of rewards).

A model of influence-based multi-agent reinforcement learning depicted in Figure 6.3.

In this model, a set of body agents with identical policies π acts in a common, shared environment. The ith body agent M_i in the state s_i selects an action a_i using current policy π, and then moves to the next state s_i'. The head agent observes changes resulting from the executed action and then calculates and assigns a r_i' to the agent as an evaluative feedback.

Equation (6.2) is a variation of the Q-learning update rule [21] used to update the values of the Q-function, and where learning homogeneity and parallelism are applied. Learning homogeneity refers to all agents building the same Q-function, and parallelism requires that they can do it in parallel. The following learning rule executes N times per step for each agent in parallel over single-shared Q-function:

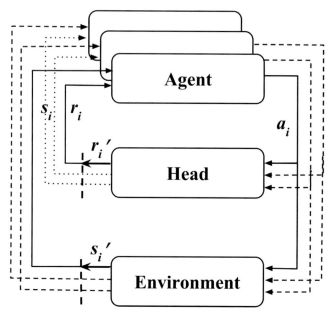

Figure 6.3 Model of Influence Based Multi-Agent Reinforcement Learning in the Case of a Holonic Homogenous Multi-Agent System.

$$\Delta Q\left(s_{i}, a_{i}\right)=\alpha\left[r_{i}^{\prime}+\gamma \max_{a \in \alpha\left(s_{i}\right)} Q\left(s_{i}, a\right)-Q\left(s_{i}, a_{i}\right)\right]. \qquad (6.2)$$

6.3 Vehicle Steering Module

The platform is based on four vehicle steering modules. The steering module consists of two wheels powered by separate motors and behaves as a differential drive. It is mounted to the platform by a bearing that allows unlimited rotation of the module with respect to the platform (Figure 6.4). The platform may be equipped with three or more modules.

The conventional approach for the platform control is a kinematics calculation and an inverse kinematics modeling [3]. The inverse kinematics calculation is known for the common schemes: the differential scheme, car scheme, and bicycle scheme. In the case of production module platforms, the four modules are controlled independently. As a consequence, the control system can only perform symmetric turning. Hence, the platform has limited maneuverability [3]. The other problem is the limitations of the robot

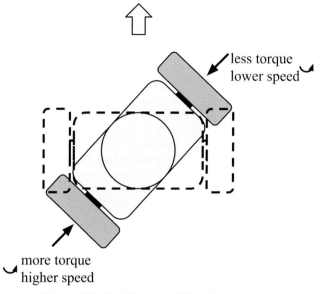

less torque
lower speed

more torque
higher speed

Figure 6.4 The Maneuverability of one module.

configuration. Previous systems require recalculations if modules are added or removed from the platform. These recalculations require a qualified engineer.

The problem of steering the robot along the trajectory is illustrated in Figure 6.5. This trajectory consists of four segments:

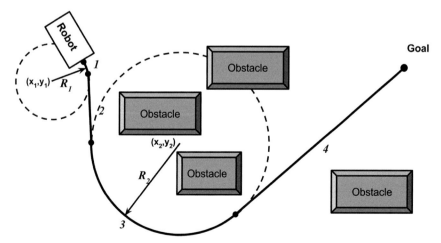

Figure 6.5 Mobile Robot Trajectory Decomposition.

- The turning radius length is R_1, the center of rotation is (x_1, y_1);
- The straight segment;
- The turning radius length is R_2, the center of rotation is (x_2, y_2);
- The straight segment.

The steering of the robot also fulfills the following specifications:

- At the starting point, the robot rotates all modules in the direction of the trajectory;
- A robot cannot stop at any point of the trajectory. The trajectory always has smooth transitions from one segment to another.

6.4 A Decomposition of Mobile Platform

A platform is composed of identical modules attached to the platform in the same way as a *multi-agent decomposition*. This is a prominent way to develop a distributed control strategy for such platforms. Mobile platforms with four identical independent driving modules can be represented as homogenous, holonic, multi-agent systems as described in Section 6.2. The driving modules are represented as body agents (or module agents) and the head agent (or platform agent) represents the whole platform. The process of multi-agent decomposition described above is shown in Figure 6.6.

The whole platform reflects global information, such as the shape of the platform and the required module topology, including its desired positions relative to the centroid of the platform. To highlight this information, we can attach a *virtual coordinate frame* to the centroid of the platform to create the virtual structure.

Figure 6.7 shows an illustrative example of the virtual structure approach with a formation composed of four vehicles capable of planar motions, where

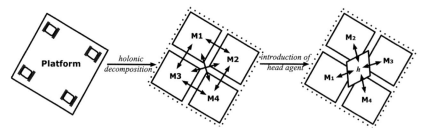

Figure 6.6 Holonic Decomposition of the Mobile Platform. Dashed lines represent the boundary of a Multi-Agent System (the Holon). Introduction of the Head Agent Leads to a reduction of communication costs.

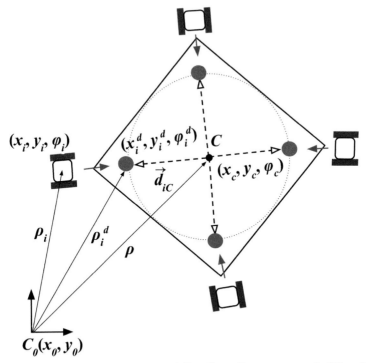

Figure 6.7   Virtual Structure with a Virtual Coordinate Frame composed of Four Modules with a known Virtual Center.

C_0 represents the beacon frame and C represents a virtual coordinate frame located at a virtual center (x_{vc}, y_{vc}) with an orientation φ_{vc} relative to C_0. Values of $\rho_i = [x_i, y_i]^T$ and $\rho_i^d = [x_i^d, y_i^d]$ represent, respectively, the -th vehicle's actual and desired position. Values of φ_i and φ_i^d represent the actual and desired orientation, respectively, of the -th vehicle. Each module's desired position $(x_i^d, y_i^d, \varphi_i^d)$ can be defined relative to the virtual coordinate frame.

For a formation stabilization with a static formation centroid, if each vehicle in a group can reach a consensus on the center point of the desired formation and specify a corresponding desired deviation from the center point, then the desired motion can be achieved [22]. If each vehicle can track its desired position accurately, then the desired formation shape can be preserved accurately.

The vectors $\vec{d_i} = (x_c - x_i, y_c - y_i)$ and $\vec{d_{iC}} = (x_c - x_i^d, x_c - y_i^d)$ represent, respectively, the -th vehicle's desired and actual deviation relative to C. The deviation vector $\vec{d_i^{err}}$ of the i-th module relative to the desired position is defined as:

$$\vec{d}_i^{err} = \vec{d}_i - \vec{d}_{iC}. \tag{6.3}$$

Each module's desired position can be defined relative to the virtual coordinate frame. Once the desired dynamics of the virtual structure are defined, the desired motion for each agent can be derived. As a result, path planning and trajectory generation techniques can be employed for the centroid while trajectory tracking strategies can be automatically derived for each module [23].

6.5 The Robot Control System Learning

The main goal of the control system is to provide the movement of the robot along the desired circular trajectory. The objective is to create a cooperative control strategy for any configuration of N modules so that all the modules within the platform achieve circular motion around the beacon. The circular motions should have a prescribed radius of rotation ρ_C defined by the center of the platform and the distance between neighbors. Further requirements are that module positioning before movement must be taken into account, and the adaptation of angular and linear speed during circular movement to reach optimal values.

We divide the process of learning into two steps:

- *Module positioning* – a learning of the module to rotate to the trajectory direction (6.5.1);
- *Cooperative movement* - a learning of cooperative motion of modules within platform (6.5.2).

The overall control architecture is depicted in Figure 6.8.

From this decomposition, every module agent will have two control policies, π_v and π_ω , for both forward and angular velocity, respectively. Policy π_ω is responsible for correct module orientation around the beacon. Each module follows this policy before the platform starts moving. Policy π_v is used during circular motion of the platform along curves. Both policies are created via reinforcement learning, which allows for generalization.

In the simulation phase, the head agent interacts with the modeling environment. In experiments with real robots, the head agent interacts with the planning subsystem. The Environment/Planning subsystem provides information about the desired speed of the platform v^d and the global state of the multi-agent $S = \left\{ U_{i=1}^N s_i \right\} \cup \{s_h\}$, where $s_i \in s$ is the state of the i-th module

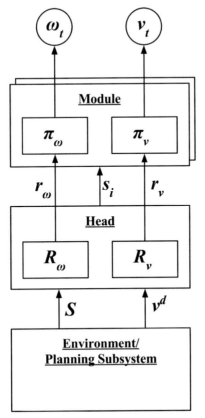

Figure 6.8 A unified view of the control architecture for a Mobile Platform.

is defined by values in Table 6.1 and $s_h = \langle C, C_0, \{U_{i=1}^N (x_i^d, y_i^d, \varphi_i^d)\}\rangle$ is the state of the head agent describes the virtual coordinate frame.

6.5.1 Learning of the Turning of a Module-Agent

This subsection describes the model for producing an efficient control rule for the positioning of a module, based on the relative position of the module with respect to the beacon. This control rule can be used for every module, since every steering module agent is homogenous.

The agent stays in a physical, 2-D environment with a reference beacon, as shown in Figure 6.9. The beacon position is defined by coordinates (x_0, y_0). The rotation radius ρ is the distance from the center of the module to the beacon.

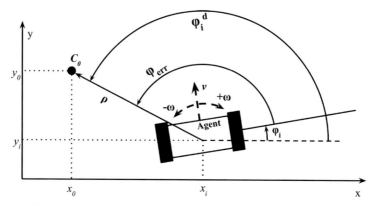

Figure 6.9   State of the Module with Respect to Reference Beacon.

The angle error is calculated using the following equations:

$$\phi_c = \arctan 2(x_0 - x_i, y_0 - y_i), \tag{6.4}$$

$$\varphi_{err} = \varphi_i^d - \varphi_i. \tag{6.5}$$

Here, φ_i^d and φ_i are known from the environment.

In the simulated model environment, all necessary information about an agent and a beacon is provided. In a real robotic environment, this information is taken from wheel odometers and a module angle sensor. The environment information states are illustrated in Table 6.1.

The full set of actions available to the agent is presented in Table 6.2. The agent with actions $A = \{A_\omega, A_v\}$ can change an angle speed by actions $A_\omega = \{\omega_+, \omega_-\}$ and linear speed by actions $A_v = \{v_+, v_-\}$. To turn, an agent controls the angular speed A_ω .

Table 6.1   The Environment Information

No	Robot Get	Value
1	X robot position, x	Coordinate, m
2	Y robot position, y	Coordinate, m
3	X of beacon center, x_b	Coordinate, m
4	Y of beacon center, y_b	Coordinate, m
5	Robot orientation angle, φ_i	Float number, radians
		$-\pi < \varphi_i \leq \pi$
6	Desired orientation angle relative to robot, φ_i^d	Float number, radians
		$-\pi < \varphi_i^d \leq \pi$
7	The radius size, r	Float number, m
8	The desired radius size, r^d	Float number, m

Table 6.2 Agent Actions

No	Robot Action	Value
1	Increase force, $v-$	+0.1, m/s
2	Reduce force, $v+$	–0.1, m/s
3	Increase turning left, $\omega+$	+0.1, rad/s
4	Increase turning right, $\omega-$	–0.1, rad/s
5	Do nothing, $\emptyset$	+0 m/s, +0 rad/s

The learning system is given a positive reward when the robot orientation is closer to the goal orientation ($\varphi_{err} \to 0$) and is using optimal speed ω_{opt}. A penalty is received when the orientation of the robot deviates from the goal orientation or the selected action is not optimal for the given position. The value of the reward is defined as:

$$r'_\omega = R_\omega(\phi^t_{err}, \omega^t). \tag{6.6}$$

Where R_ω – is a reward function, which is represented by the decision tree depicted in Figure 6.10. Here, φ_{stop} represents the value of the angle, where

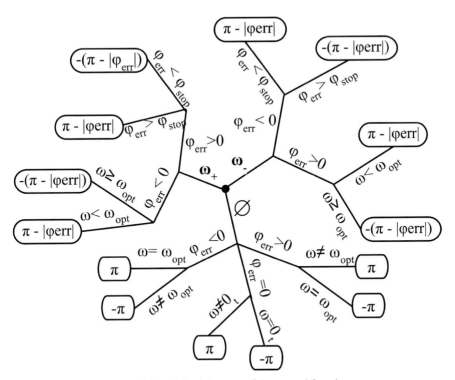

Figure 6.10   A Decision tree of the reward function.

the robot reduces speed to stop at the correct orientation, ω_{opt} [0.6 .. 0.8] rad/s, which is the optimal speed to minimize module power consumption. The parameter φ_{stop} is used to decrease the search space for the agent. When the agent angle error becomes smaller than φ_{stop}, an action that reduces the speed will receive the highest reward. The parameter ω_{opt} shows the possibility of power optimization by setting a value function. If the agent angle error is more than φ_{stop} and $\omega_{opt}^{min} < \omega < \omega_{opt}^{max}$, then the agent reward will increase. This coefficient which determines the increase ranges between [0 .. 1]. The optimization allows the use of the preferred speed with the lowest power consumption.

6.5.1.1 Simulation

The first task of the robot control is becoming familiar with robot positioning through simulation. This step is done once for an individual module before any cooperative simulation sessions. The learned policy is stored and copied for other modules via knowledge transfer. The topology of the Q-function trained during 720 epochs is shown in Figure 6.11.

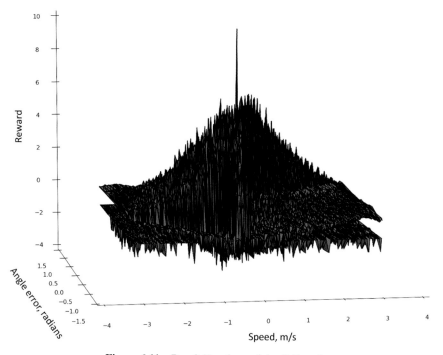

Figure 6.11   Result Topology of the Q-Function.

The external parameters of a simulation are:

- Learning rate $\alpha = 0,4$;
- Discount factor $\gamma = 0,7$;
- Minimal optimal speed $\omega_{opt}^{min} = 0,6$ rad/s;
- Maximum optimal speed $\omega_{opt}^{max} = 0,8$ rad/s;
- Stop angle, $\varphi_{stop} = 0,16$ radians.

Figure 6.12 shows the platform's initial state (left) and the positioning auto-adjustment (right) using learned policy [23].

6.5.1.2 Verification

The learning of the agent was executed on the real robot after a simulation with the same external parameters. The learning process took 1440 iterations. A real learning process takes more iterations on average because the real system has noise and sensor errors. Figure 6.13 illustrates the result of execution of a studied control system used to turn modules to the center, which is on the rear right side of the images [24].

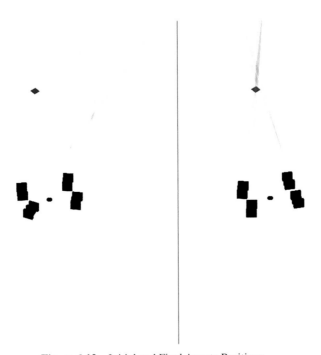

Figure 6.12   Initial and Final Agente Positions.

Figure 6.13   Execution of a Learned Control System to turn modules to the center, which is placed on the rear right relative to the platform.

6.5.2 Learning of the Turning of a Module-Agent

This subsection describes multi-agent learning for producing an efficient control law in the case of cooperative motion using an individual module's speed. The module's desired linear speed A_v should be derived through the learning process relative to the head agent so that the whole platform is moved in a circular motion.

Let the state of the module be represented by $s_t = \{v_t, \vec{d}_i^{err}\}$, where v_t is the current value of linear speed, and $\vec{d}_i^{err}$ is the error vector calculated by (6.7). Action set $A_v = \{\varnothing, v_+, v_-\}$ is represented by the increasing/decreasing of the linear speed from Table 6.2 and action $a_t \in A_v$ is a change of forward speed Δv^t for given moment in time t.

The virtual agent receives error information for each module and calculates the displacement error. This error can be positive (module ahead of the platform) or negative (module behind of the platform). The learning process progresses toward the minimization of error $\vec{d}_i^{err}$ for every module. The maximum reward is given for the case where $\vec{d}_i^{err} \rightarrow 0$, and a penalty is given when the position of the module deviates from the predefined position.

The value of the reward is defined as:

$$r_v' = \begin{cases} 1, & \vec{d}_{i\ err}' > \vec{d}_{i\ err}^t \\ -1, & \vec{d}_{i\ err}' < \vec{d}_{i\ err}^t \\ 10, & \vec{d}_{i\ err}^t = 0 \end{cases} . \qquad (6.7)$$

6.5.2.1 Simulation
Figure 6.14 shows the experimental results of the cooperative movement after learning positioning [23]. It takes 11000 epochs on average. The external parameters of a simulation are:

- Learning rate $\alpha = 0{,}4$;
- Discount factor $\gamma = 0{,}7$.

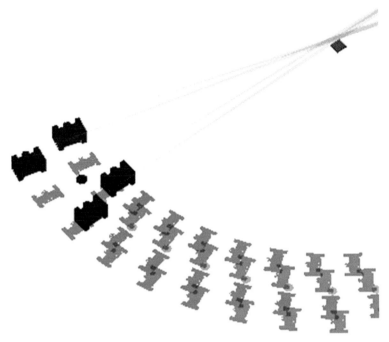

Figure 6.14   Agents Team Driving Process.

During the modules learning, the control system did not use any stabilization of the driving direction. This is because a virtual environment has an ideal, flat surface. In the case of the real platform, stabilization will be provided by internal controllers of the low-level module software. This allows us to consider only the linear speed control.

6.5.2.2 Verification

The knowledge base of the learned agents was transferred to the agents of the control system on the real robot. Figure 6.15 demonstrates the process of the platform moving by the learned system [25]. At first, modules turn in the driving direction relative to the center of rotation (the circle drawn on white paper), as shown in screenshots 1–6 in Figure 6.15. Then, the platform starts driving around the center of rotation in screenshots 7–9 in Figure 6.15. The stabilization of the real module orientation is based on a low-level controller with feedback. This controller is provided by the software control system of the robot. It helps to restrict the intellectual control system by manipulating the linear speed of modules.

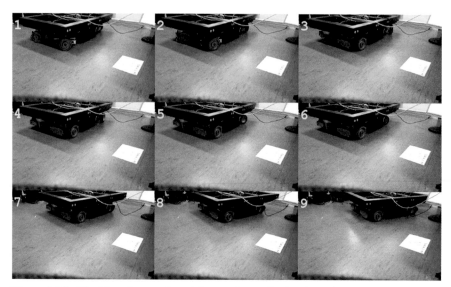

Figure 6.15 The Experiment of modules turning as in the Car Kinematics Scheme (1–6 screenshots) and movement around a White Beacon (7–9).

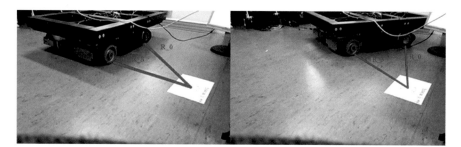

Figure 6.16 The Experiment shows that the radius doesn't change during movement.

The distance to the center of rotation is always the same on the entire trajectory of the platform. This is confirmed by Figure 6.16. Hence, the robot drives around in a circle where the coordinates of the center and the radius are known.

6.6 Conclusions

This paper focuses on an efficient, flexible, adaptive architecture for the control of a multi-wheeled, production, mobile robot. The system is based on a decomposition into a holonic, homogenous, multi-agent system and on influence-based, multi-agent reinforcement learning.

The proposed approach incorporates multiple Q-learning agents, which permits them to effectively control every module relative to the platform. The learning process was divided into two parts:

- *Module positioning* – where agents learn to minimize the error of orientation;
- *Cooperative movement* – where agents learn to adjust the desired velocity to conform to a desired position in formation.

A head agent is used to coordinate modules through the second step of learning. From this decomposition, every module agent will have a separate control policy for both forward and angular velocity.

The reward functions are designed to produce efficient control. During learning, agents take into account the current reward value and the previous reward value that helps to find the best policy of agent actions. Altogether, this provides efficient control where agents must cooperate with each other and use the policy of least resistance between each other on a real platform.

The advantages of this method are as follows:

- *Decomposition* means that instead of trying to build a global Q-function, we can build a set of local Q-functions;
- *Adaptability* – the platform will adapt its behavior for a dynamically assigned beacon and will auto-reconfigure its moving trajectory;
- *Scalability and generalization* – the same learning technique is used for every agent, for every beacon position, and for every platform configuration.

In this chapter, we showed successful experiments with the real robot where the system provides robust steering of the platform. These results indicate that the application of intellectual adaptive control systems for real mobile robots have great potential in production.

In future works, we will consider a comparison of the developed approach to mobile robot steering with existing approaches and will provide further information about efficiency of the developed control systems relative to real control systems.

References

[1] J. C. Andreas, 'Energy-Efficient ElectricMotors, Revised and Expanded', CRC Press, 1992.
[2] A. T. de Almeida, P. Bertoldi and W. Leonhard, 'Energy efficiency improvements in electric motors and drives', Springer Berlin, 1997.

[3] R. Stetter, P. Ziemniak and A. Paczynski, 'Development, Realization and Control of a Mobile Robot', In Research and Education in Robotics-EUROBOT 2010, Springer, 2011:130–140.

[4] U. Dziomin, A. Kabysh, V. Golovko and R. Stetter, 'A multi-agent reinforcement learning approach for the efficient control of mobile robot', In Intelligent Data Acquisition and Advanced Computing Systems (IDAACS), 2013 IEEE 7th International Conference on, 2, 2013:867–873.

[5] Y. Mei, Y.-H. Lu, Y. C. Hu, and C. G. Lee, 'Energy-efficient motion planning for mobile robots', In Robotics and Automation, 2004. Proceedings. ICRA'04. 2004 IEEE International Conference on, 5, 2004:4344–4349.

[6] S. Ogunniyi and M. S. Tsoeu, 'Q-learning based energy efficient path planning using weights', In proceedings of the 24th symposium of the Pattern Recognition association of South Africa, 2013:76–82.

[7] Y. Mei, Y.-H. Lu, C. G. Lee and Y. C. Hu, 'Energy-efficient mobile robot exploration', In Robotics and Automation, 2006. ICRA 2006. Proceedings 2006 IEEE International Conference on, 2006: 505–511.

[8] N. Ceccarelli, M. Di Marco, A. Garulli and A. Giannitrapani, 'Collective circular motion of multi-vehicle systems with sensory limitations', In Decision and Control, 2005 and 2005 European Control Conference. CDC-ECC'05. 44th IEEE Conference on, 2005:740–745.

[9] N. Ceccarelli, M. Di Marco, A. Garulli, and A. Giannitrapani, 'Collective circular motion of multi-vehicle systems', Automatica, 44(12): 3025–3035, 2008.

[10] D. Benedettelli, N. Ceccarelli, A. Garulli and A. Giannitrapani, 'Experimental validation of collective circular motion for nonholonomic multi-vehicle systems', Robotics and Autonomous Systems, 58(8):1028–1036, 2010.

[11] K. Fischer, M. Schillo and J. Siekmann, 'Holonic multiagent systems: A foundation for the organisation of multiagent systems', In Holonic and Multi-Agent Systems for Manufacturing., Springer, 2003: 71–80.

[12] N. Vlassis, 'A concise introduction to multiagent systems and distributed artificial intelligence', Synthesis Lectures on Artificial Intelligence and Machine Learning, 1(1):1–71, 2007.

[13] L. Gasser, 'Social conceptions of knowledge and action: DAI foundations and open systems semantics', Artificial intelligence, 47(1): 107–138, 1991.

[14] C. Gerber, J. Siekmann and G. Vierke, 'Flexible autonomy in holonic agent systems', In Proceedings of the 1999 AAAI Spring Symposium on Agents with Adjustable Autonomy, 1999.

[15] C. Gerber, J. Siekmann and G. Vierke, 'Holonic multi-agent systems', Tech. rep. DFKI Deutsches Forschungszentrum fr Knstliche Intelligenz, Postfach 151141, 66041 Saarb.

[16] C. Castelfranchi, 'Commitments: From Individual Intentions to Groups and Organizations', In *ICMAS*, 95, 1995:41–48.

[17] P. Stone and M. Veloso, 'Multiagent Systems: A Survey from a Machine Learning Perspective', Autonomous Robots, 8(3):345–383, 2000. [Online]. http://dx.doi.org/10.1023/A%3A1008942012299

[18] A. Kabysh and V. Golovko, 'General model for organizing interactions in multi-agent systems', International Journal of Computing, 11(3): 224–233, 2012.

[19] A. Kabysh, V. Golovko and A. Lipnickas, 'Influence Learning for Multi-Agent Systems Based on Reinforcement Learning', International Journal of Computing, 11(1):39–44, 2012.

[20] A. Kabysh, V. Golovko and K. Madani, 'Influence model and reinforcement learning for multi agent coordination', Journal of Qafqaz University, Mathematics and Computer Science, 33:58–64, 2012.

[21] A. G. Barto, 'Reinforcement learning: An introduction', MIT press, 1998.

[22] W. Ren and N. Sorensen, 'Distributed coordination architecture for multi-robot formation control', Robotics and Autonomous Systems, 56(4):324–333, 2008.

[23] [Online]. https://www.youtube.com/watch?v=MSweNcIOJYg

[24] [Online]. http://youtu.be/RCO-j32-ryg

[25] [Online]. http://youtu.be/pwgmdAfGb40

7

Underwater Robot Intelligent Control Based on Multilayer Neural Network

D. A. Oskin[1], A. A. Dyda[1], S. Longhi[2] and A. Monteriù[2]

[1]Department of Information Control Systems, Far Eastern Federal University
Vladivostok, Russia
[2]Dipartimento di Ingegneria dell'Informazione, Università Politecnica delle
Marche, Ancona, Italy
Corresponding author: D. A. Oskin <daoskin@mail.ru>

Abstract

The chapter is devoted to the design of an intelligent neural network-based
control system for underwater robots. A new algorithm for intelligent controller learning is derived using the speed gradient method. The proposed
systems provide robot dynamics close to the reference ones. Simulation
results of neural network control systems for underwater robot dynamics with
parameter and partial structural uncertainty have confirmed the perspectives
and effectiveness of the developed approach.

Keywords: Underwater robot, control, uncertain dynamics, multilayer
neural network speed gradient method.

7.1 Introduction

Underwater Robots (URs) promise great perspectives and have a broad
scope of applications in the area of ocean exploration and exploitation. To
provide exact movement along a prescribed space trajectory, URs need a
high-quality control system. It is well known that URs can be considered as
multi-dimensional nonlinear and uncertain controllable objects. Hence, the
design procedure of URs control laws is a difficult and complex problem
[3, 8].

Advances in Intelligent Robotics and Collaborative Automation, 147–166.

Modern control theory has derived a lot of methods and approaches to solve appropriate synthesis problems such as nonlinear feedback linearization, adaptive control, robust control, variable structure systems, etc [1, 4]. However, most of these methods for control systems synthesis essentially use information about the structure of the URs mathematical model. The nature of the interaction of a robot with the water environment is so complicated that it is hard to get the exact equations of URs motion. A possible way to overcome control laws synthesis problems can be found in the class of artificial intelligence systems, in particular, based on multi-layer Neural Networks (NNs) [1, 2, 5].

Recently, a lot of publications were devoted to the problems of NNs identification and control, starting from the basic paper [5]. Many papers are associated, in particular, with applications of NNs to the problems of URs control [1, 2, 7].

Conventional applications of multi-layer NNs are based on preliminary network learning. As a rule, this process is the minimization of the criterion which expresses overall deviations of NN outputs from the desirable values, with given NN inputs. The network learning results in NN weight coefficients adjustment. Such an approach supposes the knowledge of teaching input-output pairs [5, 7].

The feature of NNs application as a controller consists in the fact that a desirable control signal is unknown in advance. The desired trajectory (program signal) can be defined only for the whole control system [1, 2].

Thus, the multi-layer NNs application in control tasks demands a development of approaches that take into account the dynamical nature of controllable objects.

In this chapter, an intelligent NNs-based control system for URs is designed. A new learning algorithm for an intelligent NN controller, which uses the speed gradient method [4], is proposed. Numerical experiments with control systems containing the proposed NN controller were carried out in different scenarios: varying parameters and different expressions for viscous torques and forces. Modeling results are given and discussed.

Note that the choice of a NN regulator is connected with the principal orientation of the neural network approach to a priori uncertainty, which characterizes any UR. In fact, matrices of inertia of the UR's rigid body are not exactly known, as well as the added water mass. Forces and torques of viscous friction are unknown and uncertain functional structure parameters. Hence, an UR can be considered as a controllable object with partial parameter and structure uncertainties.

7.2 Underwater Robot Model

The UR mathematical model traditionally consists of differential equations describing its kinematics

$$\dot{q}_1 = J(q_1)q_2 \tag{7.1}$$

and its dynamics

$$D(q_1)\dot{q}_2 + B(q_1, q_2)q_2 + G(q_1, q_2) = U, \tag{7.2}$$

where $J(q_1)$ is the kinematical matrix; q_1, q_2 are the vectors of generalized coordinates and body-fixed frame velocities of the UR; U is the control forces and torques vector; D is the inertia matrix taking into account added masses of water; B is the Coriolis–centripetal term matrix; G is the vector of generalized gravity, buoyancy and nonlinear damping forces/torques [3].

The lack a priori knowledge of the mathematical structure and the parameters of the UR model matrices and the UR model vectors can be compensated by an intensive experimental research. As a rule, this way is too expensive and takes a long time. One alternative approach is connected with the usage of the intelligent NN control.

7.3 Intelligent NN Controller and Learning Algorithm Derivation

Our objective is to synthesize an underwater robot NN controller in order to provide the UR movement along a prescribed trajectory $q_{d1}(t)$, $q_{d2}(t)$.

Firstly, we consider the control task with respect to the velocities $q_{d2}(t)$. Let us define the error as:

$$e_2 = q_{d2} - q_2 \tag{7.3}$$

and let's introduce the function Q as a measure of the difference between desired and real trajectories:

$$Q = \frac{1}{2}e_2^T D e_2, \tag{7.4}$$

where the matrix of inertia is $D > 0$.

Furthermore, we use the speed gradient method developed by A. Fradkov [4]. According to this method, let compute the time derivative of Q:

$$\dot{Q} = e_2^T D \dot{e}_2 + \frac{1}{2}e_2^T \dot{D} e_2. \tag{7.5}$$

From

$$q_2 = q_{d2} - e_2 \tag{7.6}$$

and one has

$$D(q_1)\dot{q}_2 = D(q_1)\dot{q}_{d2} - D(q_1)\dot{e}_2. \tag{7.7}$$

Using the first term of the dynamics Equation (7.2), one can get the following:

$$D(q_1)\dot{e}_2 = D(q_1)\dot{q}_{d2} + B(q_1, q_2)q_{d2} - B(q_1, q_2)e_2 + G(q_1, q_2) - U, \tag{7.8}$$

and thus the time derivative of function Q can be written in the following form:

$$\dot{Q} = e_2^T(D(q_1)\dot{q}_{d2} + B(q_1, q_2)q_{d2} - B(q_1, q_2)e_2 + G(q_1, q_2) - U) + \frac{1}{2}e_2^T \dot{D}e_2. \tag{7.9}$$

After mathematical manipulation, one gets

$$\dot{Q} = e_2^T(D(q_1)\dot{q}_{d2} + B(q_1, q_2)q_{d2} + G(q_1, q_2) - U) - e_2^T B(q_1, q_2)e_2 + \frac{1}{2}e_2^T \dot{D}(q_1)e_2 =$$

$$= e_2^T(D(q_1)\dot{q}_{d2} + B(q_1, q_2)q_{d2} + G(q_1, q_2) - U) +$$

$$+ e_2^T(\frac{1}{2}\dot{D}(q_1) - B(q_1, q_2)e_2).$$

As known, there is a skew-symmetric matrix in the last term, hence, this term is equal to zero, and we obtain the following simplified expression:

$$\dot{Q} = e_2^T(D(q_1)\dot{q}_{d2} + B(q_1, q_2)q_{d2} + G(q_1, q_2) - U). \tag{7.10}$$

Our aim is to implement an intelligent UR control [1] based on neural networks. Without loss of generality of the proposed approach, let's choose a two-layer NN (Figure 7.1). Let the hidden and output layers have H and m neurons, respectively (m is equal to the dimension of e_2). For the sake of simplicity, one supposes that only the sum of weighted signals (without nonlinear transformation) is realized in the neural network output layer. The input vector has N coordinates.

Let's define w_{ij} as the weight coefficient for the i-th input of the j-th neuron of the hidden layer. So, these coefficients compose the following matrix

$$w = \begin{bmatrix} w_{11} & w_{12} & \cdots & w_{1N} \\ w_{21} & w_{22} & \cdots & w_{2N} \\ \cdots & \cdots & \cdots & \cdots \\ w_{H1} & w_{H2} & \cdots & w_{HN} \end{bmatrix}. \tag{7.11}$$

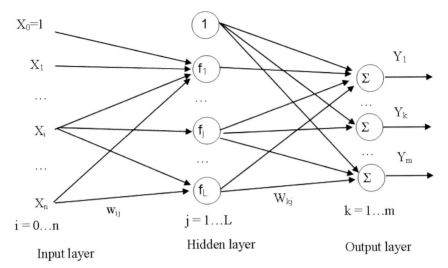

Figure 7.1   Neural network structure.

As a result of the nonlinear transformation $f(w, x)$, the hidden layer output vector can be written in the following form:

$$f(w, x) = \begin{bmatrix} f_1(w_1^T x) \\ \dots \\ f_H(w_H^T x) \end{bmatrix}, \qquad (7.12)$$

where w_k denotes the k-th raw of matrix w and x is the NN input vector.

Analogously, let's introduce the matrix W whose element W_{li} denotes the transform (weight) coefficient from the i-th neuron of the hidden layer to the l-th neuron of the output layer.

Once the NN parameters are defined, the underwater robot control signal (NN output) is computed as follows:

$$U = y(W, w, x) = W \cdot f(w, x). \qquad (7.13)$$

Substitution of this control into (7.10), allows us to get

$$\dot{Q} = e_2^T (D(q_1)\dot{q}_{d2} + B(q_1, q_2)q_{d2} + \\ + G(q_1, q_2) - W \cdot f(w, x)). \qquad (7.14)$$

To derive the NN learning algorithm, we apply the speed gradient method [4]. For this, we compute the partial derivatives of the time derivative of function Q with respect to the adjustable NN parameters – matrices and W. Direct differentiation gives

$$\frac{\partial \dot{Q}}{\partial W} = -e_2 f^T(w, x). \tag{7.15}$$

It is easy to demonstrate that if we choose all activation functions in the usual form

$$f(x) = 1/(1 + e^{-\tau x}), \tag{7.16}$$

this implies the following property

$$\frac{\partial}{\partial w_{ij}} f_i(w_i^T x) = f_i(w_i^T x)[1 - f_i(w_i^T x)]x_j. \tag{7.17}$$

Let's introduce the following additional functions

$$\phi_i(w_i^T x) = f_i(w_i^T x)[1 - f_i(w_i^T x)] \tag{7.18}$$

and the matrix

$$\Phi(w, x) = diag(\phi_1(w_1^T x)...\phi_H(w_H^T x)). \tag{7.19}$$

Hence, direct calculation gives

$$\frac{\partial \dot{Q}}{\partial w} = -\Phi W^T e_2 x^T. \tag{7.20}$$

As a final stage, one can write the NN learning algorithm in the following form:

$$W^{(k+1)} = W^{(k)} + \gamma e_2 f^T(w, x),$$
$$w^{(k+1)} = w^{(k)} + \gamma \Phi W^T e_2 x^T, \tag{7.21}$$

where γ is the learning step, k is the number of iterations.

The continuous form of this learning algorithm can be presented as

$$\dot{W} = \gamma e_2 f^T(w.x),$$
$$\dot{w} = \gamma \Phi W e_2 x^T(w.x). \tag{7.22}$$

Such an integral law of the NN-regulator learning algorithm may cause unstable regimes in the control system, as it takes place in adaptive systems [4]. The following robust form of the same algorithm is also used:

$$\dot{W} = \gamma e_2 f^T(w.x) - \alpha W,$$
$$\dot{w} = \gamma \Phi W e_2 x^T(w.x) - \alpha w, \tag{7.23}$$

where constant $\alpha > 0$.

Now, let's consider which components should be included in the NN input vector. The NN controller is oriented to compensate an influence of the appropriate matrix and vector functions, and thus, in the most common case, the NN input vector must be composed of q_1, q_2, e_2, q_{d2} and their time derivative.

The NN learning procedure leads to the reduction of function Q, and thus, in ideal conditions, the error e_2 converges to zero and the UR follows the desired trajectory

$$q_2(t) \rightarrow q_{d2}(t). \tag{7.24}$$

If the UR trajectory is given by $q_{d1}(t)$, one can choose

$$q_{d2}(t) = J^{-1}(q_1)(\dot{q}_{d1}(t) + k(q_{d1}(t) - q_1(t))), \tag{7.25}$$

where k is a positive constant. From the kinematics Equation (7.1), it follows that

$$\dot{q}_1(t) \rightarrow \dot{q}_{d1}(t) + k(q_{d1}(t) - q_1(t)) \tag{7.26}$$

and

$$\dot{e}_1(t) + k e_1(t) \rightarrow 0, \tag{7.27}$$

where

$$e_1(t) = q_{d1}(t) - q_1(t). \tag{7.28}$$

Hence, the UR follows to the planned trajectory $q_{d1}(t)$.

7.4 Simulation Results of the Intelligent NN Controller

In order to check the effectiveness of the proposed approach, different computer simulations have been carried out. The UR model parameters were taken from [6]. The UR parameters are the following:

$$D = D_{RB} + D_A,$$

where the inertia matrix of the UR rigid body is

$$D_{RB} = \begin{bmatrix} 1000 & 0 & 200 \\ 0 & 1000 & 0 \\ 200 & 0 & 11000 \end{bmatrix},$$

and the inertia matrix of the hydrodynamic added mass is

$$D_A = \begin{bmatrix} 1000 & 0 & 100 \\ 0 & 1100 & 80 \\ 100 & 80 & 9000 \end{bmatrix}.$$

Matrices B and G are

$$B = \begin{bmatrix} 210 & 20 & 30 \\ 25 & 200 & 70 \\ 15 & 33 & 150 \end{bmatrix},$$

$$G = \begin{bmatrix} 0 & 0 & 0 \end{bmatrix}^T.$$

Vector q_2 consists of the following components (linear and angular UR velocities):

$$q_2 = \begin{bmatrix} v_x & v_z & \omega_y \end{bmatrix}^T. \tag{7.29}$$

The NN input is composed by q_2 and e_2. The NN output (control forces and torque) is the vector

$$U = \begin{bmatrix} F_x & F_z & M_y \end{bmatrix}^T. \tag{7.30}$$

For the NN controller containing 10 neurons in the hidden layer, the simulation results are given on Figures 7.2 – 7.9.

In the considered numerical experiments, the desired trajectory was taken as follows:

$$\begin{cases} v_{xd} = 0.75m/\sec, \\ v_{zd} = 0.5m/\sec, \\ \omega_{yd} = -0.15rad/\sec, \end{cases} \quad 0 \le t \le 250\sec,$$

$$\begin{cases} v_{xd} = 0.5m/\sec, \\ v_{zd} = 0.75m/\sec, \\ \omega_{yd} = 0.15rad/\sec. \end{cases} \quad 250 \le t \le 500\sec$$

7.5 Modification of NN Control

In previous sections, a NN control was designed. Practically speaking, the synthesis procedure of the NN regulator does not use any information of the mathematical model of the controlled object. As one can see, differential

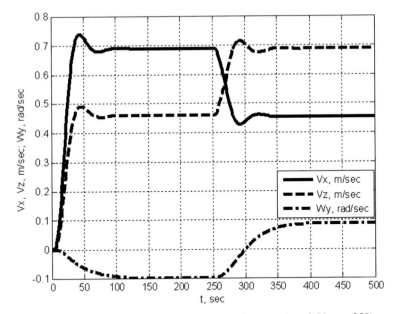

Figure 7.2 Transient processes in NN control system ($\alpha = 0.01$, $\gamma = 250$).

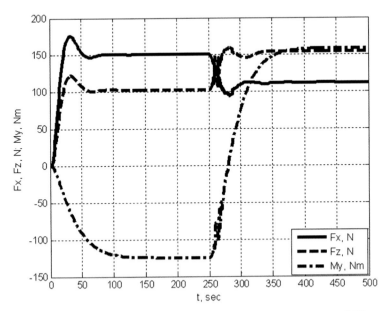

Figure 7.3 Forces and Torque in NN control system ($\alpha = 0.01$, $\gamma = 250$).

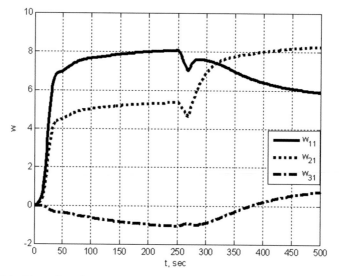

Figure 7.4   Examples of hidden layer weight coefficients evolution ($\alpha = 0.01, \gamma = 250$).

equations describing the underwater robot dynamics have a particular structure which can be taken into account for solving the synthesis problem of the control system.

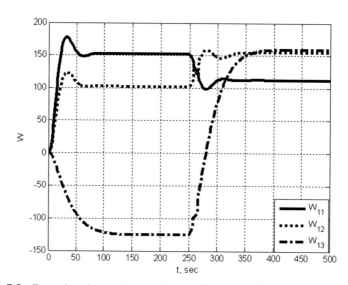

Figure 7.5   Examples of output layer weight coefficients evolution ($\alpha = 0.01$, $\gamma = 250$).

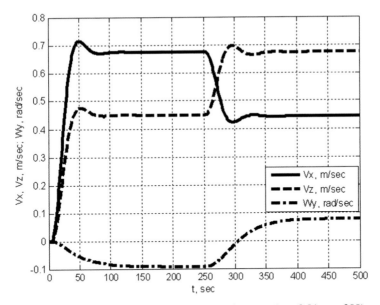

Figure 7.6 Transient processes in NN control system ($\alpha = 0.01$, $\gamma = 200$).

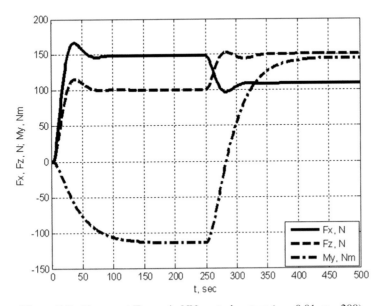

Figure 7.7 Forces ant Torque in NN control system ($\alpha = 0.01$, $\gamma = 200$).

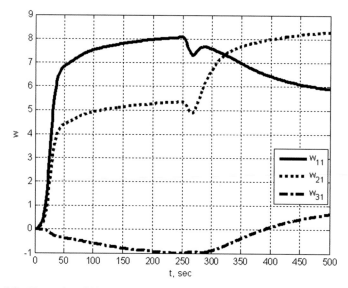

Figure 7.8   Examples of hidden layer weight coefficients evolution ($\alpha = 0.01$, $\gamma = 200$).

There exist different ways to solve it. One of the possible approaches is derived below:

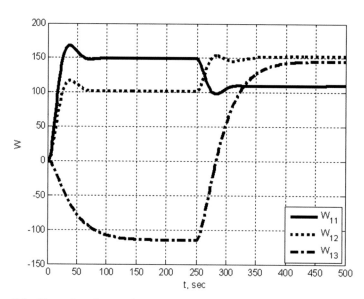

Figure 7.9   Examples of output layer weight coefficients evolution ($\alpha = 0.01$, $\gamma = 200$).

As mentioned before, the parameters of underwater robots, such as added masses, moments of inertia, coefficients of viscous friction etc, are not all exactly known because of the complex hydrodynamic nature of the robot movement in the water environment.

Let's suppose that a set of nominal UR parameters can be estimated. Hence, it is possible to get appropriate nominal matrices $D_0(q_1)$, $B_0(q_1, q_2)$ and $G_0(q_1, q_2)$ in Equation (7.2). Let's denote the deviations of the real matrices from the nominal ones as $\Delta D(q_1)$, $\Delta B(q_1, q_2)$ and $\Delta G(q_1, q_2)$, respectively. So, the following takes place:

$$\begin{aligned}
D(q_1) &= D_0(q_1) + \Delta D(q_1), \\
B(q_1, q_2) &= B_0(q_1, q_2) + \Delta B(q_1, q_2), \\
G(q_1, q_2) &= G_0(q_1, q_2) + \Delta G(q_1, q_2).
\end{aligned} \qquad (7.31)$$

Inserting expressions (7.29) into Equation (7.10) gives

$$\begin{aligned}
\dot{Q} = e_2^T (D_0(q_1)\dot{q}_{d2} + B_0(q_1, q_2)q_{d2} + G_0(q_1, q_2) + \\
+ \Delta D(q_1)\dot{q}_{d2} + \Delta B(q_1, q_2)q_{d2} + \Delta G(q_1, q_2) - U).
\end{aligned} \qquad (7.32)$$

Now let's choose the control law in the form:

$$U = U_0 + U_{NN}, \qquad (7.33)$$

where

$$U_0 = D_0(q_1)\dot{q}_{d2} + B_0(q_1, q_2)q_{d2} + G_0(q_1, q_2) + \Gamma e_2, \qquad (7.34)$$

is the nominal control associated with the known part of the robot dynamics (matrix $\Gamma > 0$ is positively definite) and U_{NN} is the neural network control to compensate the uncertainty. The scheme of the proposed NN control system for an underwater robot is given on Figure 7.10.

If the robot dynamics can be exactly determined (and uncertainty does not take place), the nominal control (7.34) fully compensates undesirable terms in (7.32) (U_{NN} can be taken as equal to zero) and one has

$$\dot{Q} = -e_2^T \Gamma e_2 < 0. \qquad (7.35)$$

Thus, functions $Q(t)$ and $e_2(t)$ converge to zero for $t \to \infty$.

In the general case, as follows from (7.32) – (7.34), one has

$$\dot{Q} = e_2^T (\Delta D(q_1)\dot{q}_{d2} + \Delta B(q_1, q_2)q_{d2} + \Delta G(q_1, q_2) - U_{NN}). \qquad (7.36)$$

As one can expect, the use of the nominal component of the control facilitates the implementation of the proper NN control.

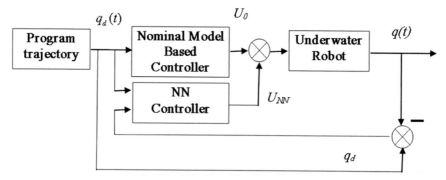

Figure 7.10 Scheme of the NN control system.

Further steps of the NN controller learning algorithm can be done practically in the same manner as above (see Equation (7.15, 20 and 21)).

In order to check the derived NN control, mathematical simulations of the UR control system were carried out. The nominal matrices $D_0\,(q_1)$, $B_0\,(q_1, q_2)$ and $G_0\,(q_1, q_2)$ were taken as follows:

$$D_0 = D_{RB0} + D_{A0},$$

$$D_{RB0} = \begin{bmatrix} 1000 & 0 & 0 \\ 0 & 1000 & 0 \\ 0 & 0 & 11000 \end{bmatrix}, D_{A0} = \begin{bmatrix} 1000 & 0 & 0 \\ 0 & 1100 & 0 \\ 0 & 0 & 9000 \end{bmatrix},$$

$$B_0 = \begin{bmatrix} 210 & 0 & 0 \\ 0 & 200 & 0 \\ 0 & 0 & 150 \end{bmatrix},$$

$$G_0 = \begin{bmatrix} 0 & 0 & 0 \end{bmatrix}^\mathrm{T}.$$

and matrix $\Gamma = diag\,[0.02,\ 0.02,\ 0.02]$.

Note that the matrices D_0, B_0 of the nominal dynamics model contain only diagonal elements which are not equal to zero. This means that the nominal model is simplified and does not take into account an interaction between different control channels (of linear and angular velocities). The absence of these terms in the nominal dynamics results in partial parametric and structural uncertainty.

Figures 7.11 – 7.18 show the transient processes and control signals (forces and torque) in the designed system with a modified NN regulator. The experimental results demonstrated that the robot coordinates converge to

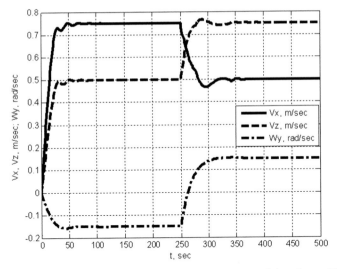

Figure 7.11   Transient processes with modified NN-control ($\alpha = 0$, $\gamma = 200$).

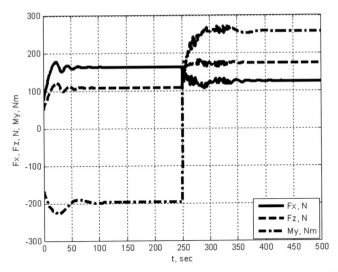

Figure 7.12   Forces and torque with modified NN control ($\alpha = 0$, $\gamma = 200$).

the desired trajectories. In comparison with the conventional multilayer NN applications, the weight coefficients of the proposed NN controller are varying simultaneously with the control processes.

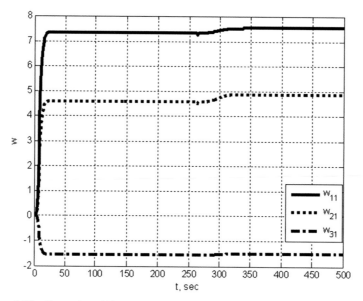

Figure 7.13   Examples of hidden layer weight coefficients evolution ($\alpha = 0$, $\gamma = 200$).

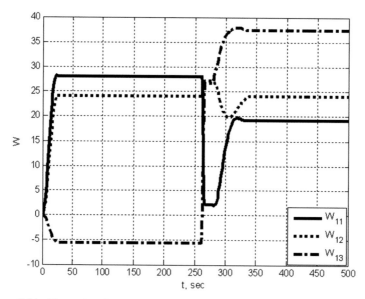

Figure 7.14   Examples of output layer weight coefficients evolution ($\alpha = 0$, $\gamma = 200$).

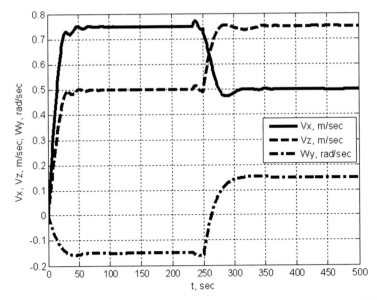

Figure 7.15 Transient processes with modified NN control (α =0.001, γ = 200).

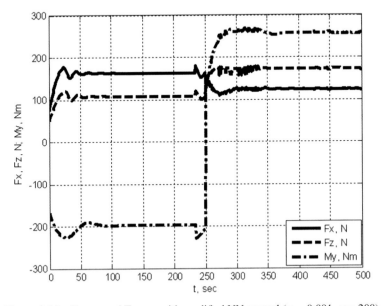

Figure 7.16 Forces and Torque with modified NN control (α = 0.001, γ = 200).

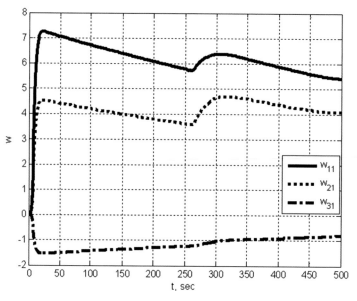

Figure 7.17   Examples of hidden layer weight coefficients evolution (α =0.001, γ =200).

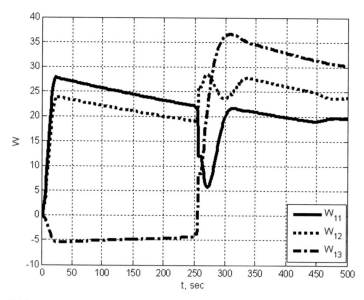

Figure 7.18   Examples of output layer weight coefficients evolution (α =0.001, γ = 200).

7.6 Conclusions

An approach on how to design an intelligent NN controller for underwater robots and how to derive its learning algorithm on the basis of a speed gradient method is proposed and studied in this chapter. The numerical experiments have shown that high-quality processes can be achieved with the proposed intelligent NN control. In the study case of producing an UR control system, the NN learning procedure allows to overcome the parameter and partial structural uncertainty of the dynamical object. The combination of the neural network approach with the proposed control, designed using the nominal model of the underwater robot dynamics, allows to simplify the control system implementation and to improve the quality of the transient processes.

Acknowledgement

The work of A.Dyda and D.Oskin was supported by Ministry of Science and Education of Russian Federation, the State Contract No 02G25.31.0025.

References

[1] A. A. Dyda, 'Adaptive and neural network control for complex dynamical objects', - Vladivostok, Dalnauka. 2007. pp. 149 (in Russian).

[2] A. A. Dyda, D. A. Oskin, 'Neural network control system for underwater robots.' IFAC conference on Control Application in Marine Systems "CAMS-2004", - Ancona, Italy, 2004, pp. 427–432.

[3] T. I. Fossen, 'Marine Control Systems: Guidance, Navigation and Control of Ships, Rigs and Underwater Vehicles', Marine Cybernetics AS, Trodheim, Norway, 2002.

[4] A. A. Fradkov, 'Adaptive control in large-scale systems', - M.: Nauka., 1990, (in Russian).

[5] A. A. Narendra, K. Parthasaraty, 'Identification and control of dynamical systems using neural networks', IEEE Identification and Control of Dynamical System, Vol.1. No 1. 20, 1990, pp. 1475–1483.

[6] A. Ross, T. Fossen and A. Johansen, 'Identification of underwater vehicle hydrodynamic coefficients using free decay tests', Preprints of Int. Conf. CAMS-2004, Ancona, Italy, 2004. pp. 363–368.

[7] R. Sutton and A. A. Craven, 'An on-line intelligent multi-input multi-output autopilot design study', Journal of Engineering for the Maritime Environment, vol. 216. No. M2, 2002, pp. 117–131.
[8] J. Yuh, 'Modeling and control of underwater robotic vehicles', Systems, Man and Cybernetics, IEEE Transactions on, vol. 20, no. 6, pp. 1475–1483, Nov/Dec 1990. doi: 10.1109/21.61218

8

Advanced Trends in Design of Slip Displacement Sensors for Intelligent Robots

Y. P. Kondratenko[1] and V. Y. Kondratenko[2]

[1]Petro Mohyla Black Sea State University, Ukraine
[2]University of Colorado Denver, USA
Corresponding author: Y. P. Kondratenko <y_kondrat2002@yahoo.com>

Abstract

The paper discusses advanced trends in design of modern tactile sensors and sensor systems for intelligent robots. The main focus is the detection of slip displacement signals corresponding to object slippage between the fingers of the robot's gripper.

It provides information on three approaches for using slip displacement signals, in particular, for the correction of the clamping force, the identification of manipulated object mass and the correction of the robot control algorithm. The study presents the analysis of different methods for the detection of slip displacement signals, as well as new sensor schemes, mathematical models and correction methods. Special attention is paid to investigations of sensors developed by the authors with capacitive, magnetic sensitive elements and automatic adjustment of clamping force. The new research results on the determination of object slippage direction based on multi-component capacity sensors are under consideration when the robot's gripper collides with the manipulated object.

Keywords: slip displacement, tactile sensor, gripper, intelligent robot, model, information processing.

Advances in Intelligent Robotics and Collaborative Automation, 167–192.

8.1 Introduction

Updated intelligent robots pose high-dynamic characteristics and effectively function under a particular set of conditions. The robot control problem is more complex in uncertain environments, as robots are usually lacking flexibility. Supplying robots with effective sensor systems provides essential extensions of their functional and technological feasibility [11]. For example, a robot may often encounter a problem of gripping and holding i object doing manipulation processes with the required clamping force F_i^r avoiding its deformation or mechanical injury, $i = 1...n$. To successfully solve the current tasks, the robots should possess the capability to recognize the objects by means of their own sensory systems. Besides, in some cases, the main parameter due to which robot can distinguish objects of the same geometric shape is their mass $m_i(i = 1...n)$. The robot sensor system should identify the mass m_i of each i-th manipulated object in order to identify a class (set) an object refers to.

The sensor system should develop the required clamping force F_i^r corresponding to mass value m_i, as $F_i^r = f(m_i)$. Such current data may be applied when the robot functions in dynamic or random environments. For example, in a case when the robot should identify unknown parameters for any type of object and location in robot's working zone. The visual sensor system may not always be utilized, in particular, in poor vision conditions. Furthermore, in cases when the robot manipulates with an object of variable mass $m_i(t)$, its sensor system should provide the appropriate change of clamping force value $F_i^r(t) = f[m_i(t)]$ for the gripper fingers. This information can also be used for the robot control algorithm correction, since the mass of the robot arm's last component and its summary inertia moment vary.

8.2 Analysis of Robot Task Solving Based on Slip Displacement Signals Detection

One of the current approaches to solving the mass m_i identification problem of grasped objects and producing the required clamping force F_i^r is in the development of tactile sensor systems based on object slippage registration [1, 11, 17, 18, 20, 22] while slipping between the gripper fingers (Figure 8.1).

As usual, the slippage signal detection in robotic systems is accomplished either in the trial motion or in the regime of continuous lifting of the robot arm. In some cases, during the process of trial motions, it is necessary to make a series of trial motions (Figure 8.2) for creating the required clamping forces

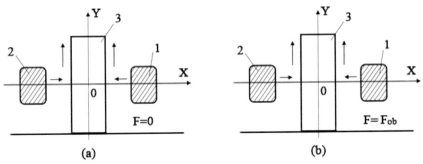

Figure 8.1   Grasping and lifting an object with the robot's arm: Initial positions of the gripper fingers (1,2) and object (3) (a); Creating the required clamping force F_{ob} by the gripper fingers during object slippage in the lifting process (b).

F_{ob} or $F_{ie} = kF_{ob}$, where k is a coefficient which impacts the reliability of the object motion (by the robot arm) in the realisation of the required path, $k > 1$.

According to Figure 8.2, the robot creates a minimal value of clamping force F_{min} in time moment t_1. Then step by step robot lifts an object vertical distance Δl and the robot gripper increases the clamping force $(F(t_1) + \Delta F)$ if a slip displacement signal appears. The grasping surface of the object is limited by value l_{max}. The first series of trial motions is finished in time moment t_2, when $l = l_{max}$ (Figure 8.1(b)). After that, the robot decompresses the fingers $(t_2...t_3)$, moves the gripper $(t_3...t_4)$ to the initial position (Figure 8.1(a)) and creates $(t_4...t_5)$ the initial value of the clamping force $F(t_5) = F(t_2) + \Delta F = F_1$ for the beginning of a second stage or second series of trial motions.

Some sensor systems based on the slip displacement sensors were considered in [24, 25], but random robot environments very often requires the development of new robot sensors and sensor systems for increasing the speed of operations, the growth of positioning accuracy or the desired path-following precision.

Thus, the task of the registration of slippage signals between the robot fingers for manipulated objects is connected with: a) the necessity of the required force creation being adequate to the object's mass value; b) the recognition of objects; c) robot control algorithm correction.

The idea of a trial motion regime comprises the process of an iterative increase in the compressive force value if the slippage signal is being detected. The continuous lifting regime provides the simultaneous object lifting process and increasing clamping force until the slippage signal disappears. The choice of the slip displacement data acquisition method depends on a robot's purpose,

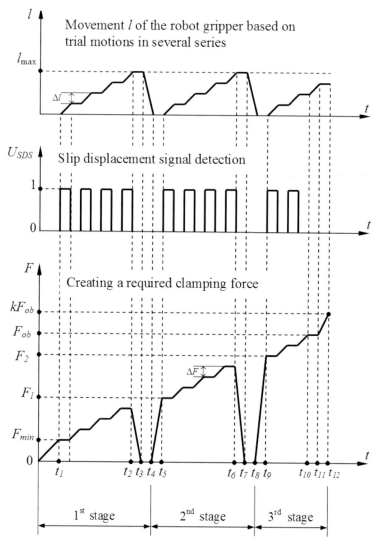

Figure 8.2   Series of trial motions with increasing clamping force F of gripper fingers based on object slippage.

the salient features of its functioning medium, the requirements of its response speed and the performance in terms of an error probability.

Figure 8.3 illustrates the main tasks in robotics which can be solved based on slip displacement signal detection.

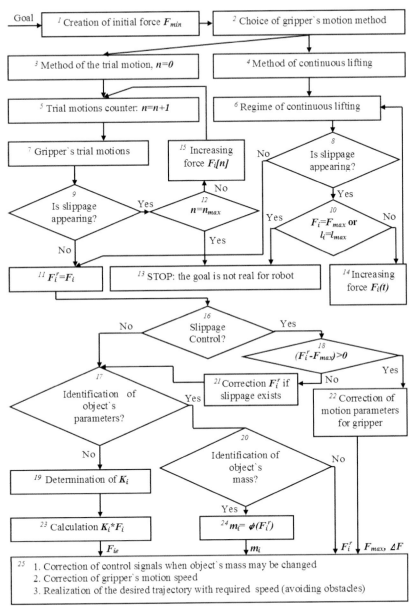

Figure 8.3 The algorithm for solving different robot tasks based on slip signal detection.

8.3 Analysis of Methods for Slip Displacement Sensors Design

Let's consider the main methods of slip displacement data acquisition, in particular [3, 4, 10, 11, 14, 17, 18, 20, 22]:

The vibration detection method. This method is based on a principle of the vibration detection in the sensing element when the object is slipping. To implement the method mentioned, the following sensing elements may be adopted: a sapphire needle interacting with the crystal receiver or a rod with a steel ball, connected to the electromagnetic vibrator.

The pressure re-distribution detection method. The method relies on the detection of a distribution change in pressure between gripper fingers at object slippage and is based on the physiological sensitivity function of human skin. The pressure transducers serve as nerves and are surrounded by an elastic substance as in the human body.

The rolling motion detection method. The method is characterized by transducing the object displacements in the vertical direction when slipping to the rolling motion of a sensitive element. A slip displacement signal is detected through rolling of a cylinder roller with elastic covering and a large friction coefficient. The roller's rolling motions may be converted into an electric signal by means of photoelectric or magnetic transducers, containing a permanent magnet on a movable roller, and in the case of a magnetic head being placed on a gripper.

The impact-sliding vibrations detection method. A core of the method implies the detection of liquid impact-sliding vibrations when the object is slipping. An acrylic disk with cylinder holes is used in the slip displacement sensor, realizing the method under consideration. A rubber gasket made in the form of a membrane protects one end of the disk, and a pressure gauge is installed on the other end. The hole is filled with water so that its pressure slightly exceeds the atmospheric pressure. While the motion of the object is in contact with a membrane, the impact-sliding vibrations are appearing and, therefore, inducing impulse changes on the water pressure imposed by a static pressure.

The acceleration detection method. This method is based on the measurement of accelerations of the sensing element motion by the absolute acceleration signal separation. The slip displacement sensor, comprising two accelerometers, can be used in this case. One of the accelerometers senses the absolute acceleration in the gripper, another responds to the acceleration of the sensitive plate springing when the detail is slipping. The sensor is attached to the computer identifying the slip displacement signal by comparing the output signals of both accelerometers.

The interference pattern change detection method. This method involves the conversion of the intensity changes reflected from the moving surface of the interference pattern. The intensity variation of the interference pattern is converted to a numerical code, the auto-correlation function is computed and it achieves its peak at the slip displacement disappearance.

The configuration change detection in the sensitive elements method. The essence of the method incorporates the measurement of the varying parameters when the elastic sensitive element configuration changes. The sensitive elements made of conductive rubber afford coating of the object surface protruding above the gripper before the trial motion. When the object is displacing from the gripper, the configuration changes, the electrical resistance of such sensitive elements changes accordingly, confirming the existence of slippage.

The data acquisition by means of the photoelastic effect method. An instance representing this method may be illustrated by a transducer, in which, under the applied effort, the deformation of sensitive leather produces the appearance of voltage in the photoelastic system. The object slippage results in the change of the sensitive leather deformation being registered by the electronic visual system. The photosensitive transducer is a device for the transformation of interference patterns into the form of a numerical signal. The obtained image is of binary character, each pixel gives one bit of information. The binary representation of each pixel enables to reduce the processing time.

The data acquisition based on friction detection method. The method ensures the detection of the moment when the friction between the gripper fingers and the object to be grasped goes from friction at rest to dynamic friction.

Method of fixing the sensitive elements on the object. The method is based on fixing the sensitive elements on the surface of the manipulated objects before the trial motions with the subsequent monitoring of their displacement relative to the gripper at slipping.

Method based on recording oscillatory circuit parameters. The method is based on a change in the oscillatory circuit inductance while the object slips. The inductive slip sensor with a mobile core, stationary excitation winding and solenoid winding being one of the oscillatory circuit branches implements the method. The core may move due to the solenoid winding. The reduction of the solenoid winding voltage indicates the process of lowering. The core is lowering under its own weight from the gripper center onto the object to be grasped. The oscillatory circuit induces the forced oscillations with the frequency coinciding with the frequency of excitation in the excitation winding.

The video signal detection method. The basis of this method constitutes a change in detection and ranging of patterns or video pictures as an indication of object slippage. The slip displacement detection is accomplished by means of the location sensors or visual sensors based on a laser source that has either a separated and reflecting beam or a vision with a non-coherent beam of light conductors for picture lighting and a coherent beam for image transmission.

The choice of a slip displacement detection method involves the multicriterion approach taking into account the complexity of implementation, the bounds of functional capabilities, mass values and overall dimensions, reliability and cost.

8.4 Mathematical Model of Magnetic Slip Displacement Sensor

8.4.1 SDS Based on "Permanent Magnet/Hall Sensor" Sensitive Element and Its Mathematical Model

In this chapter, the authors consider a few instances of updating the measurement systems. To suit the requirements of increasing the noise immunity of the vibration measurement method, a modified method has been developed. The modified method is founded on the measurement of the sensitive element angular deviation occurring at the object slippage (Figure 8.4).

Let's consider the structure and mathematical model (MM) of the SDS developed by the authors with a magnetic sensitive element which can detect the bar's angular deviation appearing at object slippage (Figure 8.4). The dependence $U = f(\alpha)$ can be used to determine the sensitivity of the SDS and the minimal possible amplitudes of the robot trial motions.

To construct the mathematical models, consider a magnetic system comprising a prismatic magnet with dimensions $c \times d \times l$, which is set to ferromagnetic plane with infinite permeability $\mu = \infty$ (Figure 8.5), where: c- width, d- length, and l- height of magnet, $(d \gg l)$. The point $P(X_P, Y_P)$ is the observation point, which is located on the vertical axis and can change its position relative to the horizontal axis Ox or vertical axis Oy.

A Hall sensor with a linear static characteristic is located at the observation point P. Let's form the mathematical model for the determination of the magnetic induction B and the output voltage $U_{out}(P)$ of the Hall sensor in

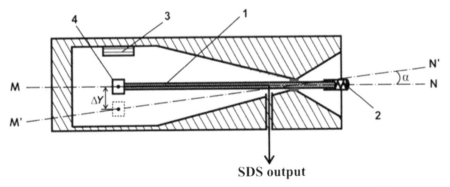

SDS output

Figure 8.4 Magnetic SDS: 1– Rod; 2– Head; 3– Permanent magnet; 4– Hall sensor.

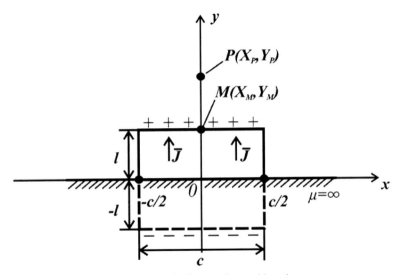

Figure 8.5 Model of magnetic sensitive element.

relation to an arbitrary position of the observation point P under the surface of the magnet.

The value of magnetic induction (outside the magnet volume) is $\bar{B} = \mu_0 \bar{H}$, where μ_0 is a magnetic constant; $\bar{H}$ is the magnetic field strength vector.

In the middle of the magnet– the magnetic induction value is determined by the dependence $\bar{B} = \mu_0(\bar{J} + \bar{H})$, where J is a magnetization value.

$$J = J_0 + \chi H,$$

where χ is a magnetic susceptibility; J_0 is the residual magnetization value.

The permanent magnet can be represented [26-28] as a simulation model of the surface magnetic charges that are evenly distributed across the magnet pole faces with the surface density J_T.

Thus, a y- component of the magnetic field strength H_y of the magnetic charges can be calculated as:

$$H_y = -\frac{J_T}{2\pi} \left[\left(\text{arctg}\frac{X_p+c/2}{l-Y_p} - \text{arctg}\frac{X_p-c/2}{l-Y_p} \right) - \left(\text{arctg}\frac{X_p+c/2}{-l-Y_p} - \text{arctg}\frac{X_p-c/2}{-l-Y_p} \right) \right],$$

(8.1)

and the y- component of magnetic induction B_y can be presented as:

$$B_y = -\frac{J_T\mu_0}{2\pi} \left[\left(\text{arctg}\frac{X_p+c/2}{l-Y_p} - \text{arctg}\frac{X_p-c/2}{l-Y_p} \right) - \left(\text{arctg}\frac{X_p+c/2}{-l-Y_p} - \text{arctg}\frac{X_p-c/2}{-l-Y_p} \right) \right],$$

To determine the parameter J_T, it is necessary to measure the induction of the pole faces center $B_y\big|_{x=0,y=l+}$. Value $(y = l+)$ indicates that the measurement of B_{mes} is conducted outside the volume of the magnet. The value of magnetic induction at a point with the same coordinates on the inside of the pole faces can be considered equal to the value of induction from the outside pole faces by virtue of the continuity of the magnetic flux and lines of magnetic induction, namely:

$$B_y\big|_{x=0,y=l+} = B_y\big|_{x=0,y=l-}.$$

So, we can write: $B_y\big|_{x=0,y=l+} = B_{mes} = \mu_0 J_t + \mu_0 H_y\big|_{x=0,y=l-}$, where B_{mes} is the value of magnetic induction measured at the geometric center of the top pole faces of the prismatic magnet.

On the basis of (8.1), we obtain:

$$B_{mes} = \mu_0 J_T \left[1 - \lim_{(Y_p \to l-)} \frac{2}{2\pi} \left(\text{arctg}\frac{c/2}{l-Y_p} + \text{arctg}\frac{c/2}{l+Y_p} \right) \right]$$

$$= \mu_0 J_T \left(\frac{1}{2} + \frac{1}{\pi}\text{arctg}\frac{c}{4l} \right),$$

$$J_T = \frac{2\pi B_{mes}}{\mu_0 \left(\pi + 2\text{arctg}\frac{c}{4l} \right)}.$$

For the y- component of the magnetic induction B_y (P) at the observation point P, the following expression was obtained:

$$B_y(p) = -\frac{B_{mes}}{\left(\pi + 2arctg\frac{c}{4l}\right)} \left[\left(arctg\frac{X_p + c/2}{l - Y_p} - arctg\frac{X_p - c/2}{l - Y_p}\right) - \left(arctg\frac{X_p + c/2}{-l - Y_p} - arctg\frac{X_p - c/2}{-l - Y_p}\right)\right]. \qquad (8.2)$$

8.4.2 Simulation Results

For the analysis of the existing mathematical model (8.2), let's calculate the value of magnetic induction on the surface of the magnet (Barium Ferrite) with parameters of $c=0.02$m, $d=0.08$m, $l=0.014$m and a value of magnetic induction $B_{mes} = 40$ mT (value measured at the geometric center of the upper limit of the magnet).

The simulation results for the magnetic induction are represented as $B_y = f_i(X_P), i = 1,2,3$ above the magnet for different values of the height Y_P of the observation point P (X_P, Y_P), where indicated:

f_1 – for

$$B_y\Big|_{x\in[-20;20]mm,y=l+1mm};$$

f_2 – for

$$B_y\Big|_{x\in[-20;20]mm,y=l+5mm};$$

f_3 – for

$$B_y\Big|_{x\in[-20;20]mm,y=l+20mm}.$$

As can be seen from Figure 8.6, the magnetic induction $B_y = f_1$ (X_P) above the surface of the magnet is practically constant for the coordinate $X_P \in [-5; 5]$ mm, which is half of the corresponding size of the magnet. If the distance between the observation point P and magnet increases (f_2 (X_P), f_3 (X_P) in Figure 8.6), the curve shape changes become more gentle, with a pronounced peak above the geometric center of the top pole faces of the prismatic magnet (at the point $X_P = 0$).

For the Hall sensor (Figure 8.4) in the general case, the dependence of the output voltage U_{out} (P) on the magnitude of the magnetic induction B_y is defined as:

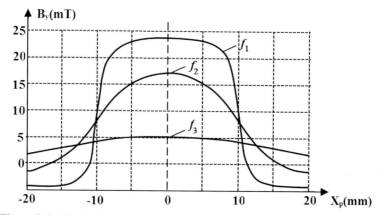

Figure 8.6   Simulation results for B_y (P) based on the mathematical model (8.2).

$$U_{out}\ (P) = U_c + kB_y(P), \qquad (8.3)$$

where k is the correction factor, that depends on the type of Hall sensor; U_c is a constant component of the Hall sensor output voltage.

For Hall sensors with linear dependence of the output signal U_{out} on the magnetic induction B_y, $k = const$ and $k = f\{B_y\ (P)\}$ for nonlinear dependence (8.3). For Hall sensor SS490 (Honeywell) with linear dependence (8.3), the values of the parameters are: $U_c = 2.5V$ and $k = 0.032$ (according to the static characteristic of the Hall sensor). The authors present the mathematical model of the Hall sensor output voltage $U_{out}\ (Y_P)$ at its vertical displacement above the geometric center of the top pole faces of magnet $(X_P = 0)$:

$$U_{out}(Y_P) = 2,5 + 7,4 \times 10^{-3} \left(arctg\frac{0,01}{0,014 - Y_P} + \right.$$
$$\left. arctg\frac{0,01}{0,014 + Y_P} \right). \qquad (8.4)$$

The comparative results for dependences $U_{out}\ (Y_p)$, $U_E\ (Y_p)$ and $U_R(Y_p)$ are presented in Figure 8.7, where $U_{out}\ (Y_p)$ was calculated using MM (8.4), $U_E\ (Y_p)$ are the experimental results according to [7] and $U_R\ (Y_p)$ is a nonlinear regressive model according to [8].

The comparative analysis (Figure 8.7) of the developed mathematical model $U_{out}\ (Y_p)$ with the experimental results $U_E\ (Y_p)$ confirms the correctness and adequacy of the synthesized models (8.1)–(8.4).

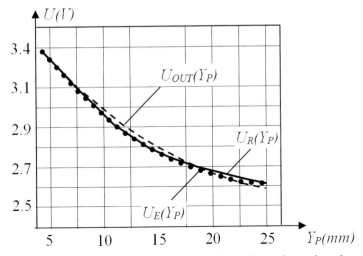

Figure 8.7 Comparative analysis of modeling and experimental results.

8.5 Advanced Approaches for Increasing the Efficiency of Slip Displacement Sensors

This study presents a number of sensors for data acquisition in real time [5, 11, 13, 14]. The need for rigid gripper orientation before a trial motion has caused the development of the slip sensor based on a cylinder roller with a load which has two degrees of freedom [3, 14].

The sensitive element of the new sensor developed by the authors has the form of a ball (Figure 8.8) with light-reflecting sections disposed in a staggered order (Figure 8.9), thus providing slippage detection by the photo-method.

The ball is arranged in the sensor's space through spring-loaded slides. Each slide is connected to the surface of the gripper's space by an elastic element made of conductive rubber.

The ball motion is secured by friction-wheels and is measured with the aid of incremental transducers in another modification of the slip sensor with the ball acting as a sensitive element. The ball contacts with the object through the hole. In this case, the ball is located in the space of compressed air dispensed through the hole.

For the detection of the sensitive element angular deviation during object slippage in any direction, the authors propose [12] a slip displacement sensor with a measurement of changeable capacitance (Figure 8.10).

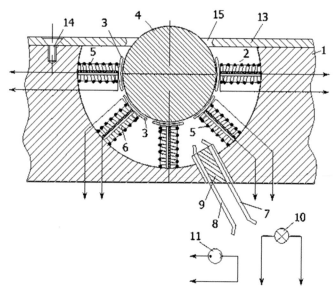

Figure 8.8   The ball as sensitive element of SDS: 1– Finger of robot's gripper; 2– Cavity for SDS instalation; 3– Guides; 4– Sensitive element (*a* ball); 5– Spring; 6– Conductive rubber; 7, 8– Fiber optic light guides; 9– *a* Sleeve; 10– Light; 11– Photodetector; 13–Cover; 14–Screw; 15–Hole.

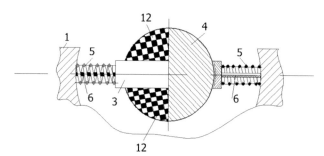

Figure 8.9   Light-reflecting surface of the sensitive ball with reflecting and absorbing portions (12) for light signal.

The structure of the intelligent sensor system developed by the authors which can detect the direction of object slippage based on the capacitated SDS (Figure 8.10) is represented in Figure 8.11 with channels of information processing and electronic units that implement the intellectual base of production rules to identify the direction of displacement of the object in the gripper (if there is a collision with an obstacle).

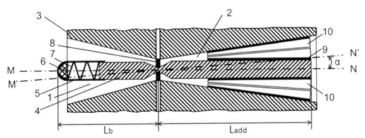

Figure 8.10 Capacitated SDS for the detection of object slippage in different directions: 1– Main cavity of robot's gripper; 2– Additional cavity; 3– Gripper's finger; 4– Rod; 5– Tip; 6– Elastic working surface; 7– Spring; 8– Resilient element; 9, 10– Capacitor plates.

The SDS is placed on at least one of the gripper fingers (Figure 8.10). The recording element consists of four capacitors distributed across the conical surface of the additional cavity (2). One plate (9) of each capacitor is located on the surface of the rod (4) and the second plate (10) is placed on the inner surface of the cavity (2).

The intelligent sensor system (Figure 8.11) provides an identification of signals corresponding to object slippage direction {*N, NE, E, SE, S, SW, W, NW*} in the gripper in the cases of contacting obstacles.

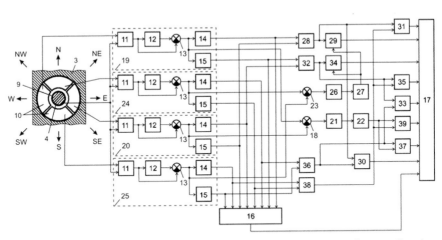

Figure 8.11 Intelligent sensor system for identification of object slippage direction: 3– Gripper's finger; 4– Rod; 9, 10– Capacitor plates; 11– Converter "capacitance-voltage"; 12– Delay element; 13, 18, 23– Adders; 14, 15, 21, 26– Threshold elements; 16– Multi-Inputs element OR; 17– Computer information-control system; 19, 20, 24, 25– Channels for sensor information processing; 22, 27– Elements NOT; 28–39– Elements AND.

The implementation of relation (Figure 8.10)

$$\frac{L_b}{L_{add}} = \frac{1}{5}$$

allows to increase the sensitivity of developed sensor system.

Initially (Figure 8.10), the tip (5) is held above the surface of the gripper's finger (3) by a spring (7), and a resilient element (8) holds a rod (4) in such a position that its longitudinal axis is perpendicular to the surface of the finger and coincides with the axis MN of the rod (4). When gripping a manipulation object, its surface comes in contact with the tip (5), the spring (7) is compressed and the tip (5) is immersed in a cavity (2). At this moment, the value of compressive force corresponds to the minimal pre-determined value F_{min} that eliminates distortion or damage of the object.

The object begins to slip in the gripper if an inequality between compressive force and the mass of the object exists during the trial motion. In this case, the rod (4) is deflecting along the sliding direction on the angle α as the result of the friction forces between the object's contacting surface and the working surface (6) of the tip (5).

Thus, the longitudinal axis of the rod (4) coincides with the axis M'N'. Reciprocal movement of plates (9) and (10) with respect to each other in all capacitative elements leads to value changes of the capacities C_1, C_2, C_3 and C_4 depending on the direction of rod's movement. The changes of the capacities lead to the voltage changes at the outputs of the respective convertors "capacitance-voltage" in all sensory processing channels.

From time to time, the robot's gripper may be faced with an obstacle when the robot moves a manipulation object in dynamic environments according to preplanned trajectories. The obstacles can appear randomly in the dynamic working area of the intelligent robot. As a result of the collision between the robot gripper and the obstacle, object's slippage may appear if the clamping force F is not enough for reliable fixation of object between the gripper's fingers. The direction {N, NE, E, SE, S, SW, W, NW} of such slip displacement of the object depends on the position of the obstacle in the desired trajectory.

In this case, the output signal of element OR (16) is equal to 1 and this is a command signal for the computer system (17) which constrains the implementation of the planned trajectory. At the same time, the logical 1 signal appears at the output of one of the AND elements {29, 30, 31, 33, 34, 35, 37, 39} that corresponds to one of the object's slippage direction {N, NE, E, SE, S, SW, W, NW} in the robot gripper.

Let's consider, for example, the determination of slippage direction $\{N\}$ after the contact between the obstacle and robot gripper (with object). If the object slips in direction $\{N\}$ (Figure 8.11), then:

- the capacity C_1 in first channel (19) increases and a logical 1 signal appears at the output of threshold element (14) and at the first input of the AND element (28);
- the capacity C_3 in the third channel (20) decreases and a logical 1 signal appears at the output of threshold element (15), at the second input and output of AND element (28) and at the first input of the AND element (29);
- the capacities C_2, C_4 of the second (24) and fourth (25) channels of the sensor information processing are equivalent $C_2 = C_4$ and in this case a logical 0 signal appears at the outputs of adder (23) and threshold element (26), a logical 1 signal appears at the output of NOT element (27), at the second input and output of AND element (29) and at the second input of the computer information-control system (17). It means that the direction of the object's slippage is $\{N\}$, taking into account that output signals of the AND elements $\{30, 31, 33, 34, 35, 37, 39\}$ equal 0.

The production rules "IF-THEN" base are represented in Table 8.1. This rule base determines the functional dependence between the direction of object slippage (Figure 8.11), the current state of each capacitor $\{C_1, C_2, C_3, C_4\}$ and the corresponding output signals $\{U_1, U_2, U_3, U_4\}$ of the multi-capacitor slip displacement sensor, where: U_i, $(i = 1...4)$– output signal of i-th converter "capacitance - voltage" (11); $(>)$– indicator of the corresponding signal $U_i,(i = 1...4)$ increases during the object slippage process; $(<)$– indicator of the corresponding signal $U_i,(i = 1...4)$ decreases during the object slippage process; $(=)$– pair's indicator of equivalence according to

Table 8.1 The base of production rules "IF-THEN" for indetification of the slip displacement direction

Number of Production Rule	Antecedent				Consequent
	U_1	U_2	U_3	U_4	Direction of Slippage
1	>	=	<	=	N
2	>	>	<	<	NE
3	=	>	=	<	E
4	<	>	>	<	SE
5	<	=	>	=	S
6	<	<	>	>	SW
7	=	<	=	>	W
8	>	<	<	>	NW

conditions $U_i = U_j$, $(i = 1...4)$, $(j = 1...4)$, $i \neq j$ in the antecedents of the production rules.

The mathematical models of different types of slip displacement sensors with a measurement of changeable capacity are presented in [5, 19, 21, 23].

8.6 Advances in Development of Smart Grippers for Intelligent Robots

8.6.1 Self-Clamping Grippers of Intelligent Robots

The slip displacement signals, which are responsible for the creation of the required compressive force adequate to the object mass, provide the conditions for the correction of the gripper trajectory-planning algorithm, which identifies an object mass as a variable parameter [9]. The object mass identification is carried out in response to the final value of the compressive force, recorded at the slippage signal disappearance. It is of extreme importance to employ slip sensors with uprated response when the object mass changes in the functioning process.

In those cases when the main task of the sensing system is the compression of the object without its deformation or damage, it is expedient in future research to project the advanced grippers of a self-clamping design (Figure 8.12), excluding the gripper drive for the compressive force growth (at slippage) up to the required value.

The information processing system of a self-adjusting gripper of an intelligent robot with angle movement of clamping rollers consists of: 24– control unit; 25– delay element; 26, 30, 36– adder; 27, 32, 37, 41– threshold element; 28, 29, 31, 33, 34, 35, 38, 40, 42– switch; 39– voltage source.

In such a gripper (Figure 8.12), the rollers have two degrees of freedom and during object slippage they have a compound behavior (rotation and translation motions). This gripper (Figure 8.12) is adaptive since object self-clamping is being accomplished with a force adequate to the object mass up to the moment of the slippage disappearance [6, 14].

Another example of a developed self-clamping gripper [15] is represented in Figure 8.13, where: 1– finger; 2– roller axis; 3– roller; 4– sector element; 5– guide gear racks; 6– pinion; 7– travel bar; 8, 9– axis; 10– object; 11– elastic working surface; 12, 13– spring; 14, 15– clamping force sensor; 16– electroconductive contacts; 17, 18– fixator; 19, 20– pintle.

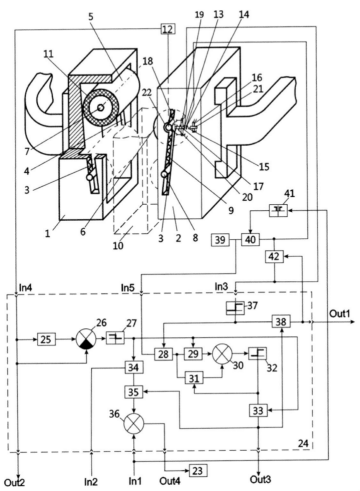

Figure 8.12 Self-adjusting gripper of an intelligent robot with angle movement of clamping rollers: 1, 2– Finger; 3, 4– Guide groove; 5, 6– Roller; 7, 8– Roller axis; 9, 15, 20– Spring; 10– Object; 11, 18– Elastic working surface; 12– Clamping force sensor; 13, 14– Electroconductive contacts; 16, 19– Fixator; 17– Stock; 21– Adjusting screw; 22– Deepening; 23– Finger's drive.

The experimental self-clamping gripper with plane-parallel displacement of the clamping rollers and intelligent robot with 4 degrees of freedom for experimental investigations of the developed grippers and slip displacement sensors are represented in Figure 8.14 and Figure 8.15, correspondingly.

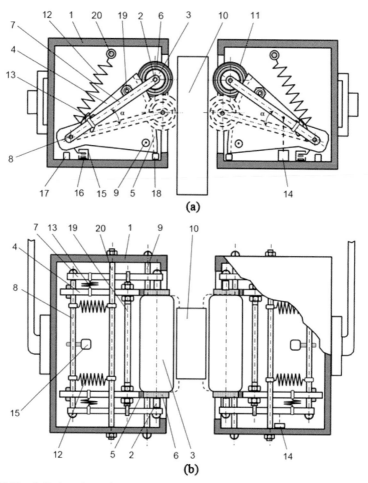

Figure 8.13   Self-clamping gripper of an intelligent robot with plane-parallel displacement of the clamping roller: Front view (a); Top view (b).

8.6.2 Slip Displacement Signal Processing in Real Time

Frequent handling operations require a compressive force being exerted through the intermediary of the robot's sensing system in the continuous hoisting operation. This regime shows a simultaneous increase in the compressive force while the continuous hoisting operation and lifting of the gripper in the vertical direction take place accompanied by the slip displacement signal measurement. When the slippage signal disappears, the compressive force does not increase and, therefore, the operations with the object are

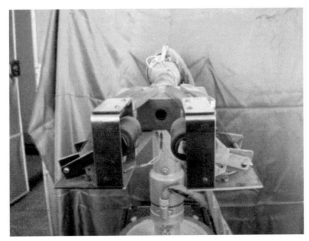

Figure 8.14 Experimental self-clamping gripper with plane-parallel displacement of the clamping rollers.

Figure 8.15 Intelligent robot with 4 degrees of freedom for experemental investigations of SDS.

accomplished according to the robot's operational algorithm. To realize the trial motion regime and the continuous hoisting operation being controlled in real time, stringent requirements to the parameters should be met, in particular:

- the response time between the moment of slippage emergence and the moment when the gripper fingers begin to increase the compressive force;

- the time of the sliding process including the moments between the emergence of sliding and its disappearance;
- the minimal object displacement detected by the slip signal.

The problem of raising the sensors response in measuring the slip displacement signals is tackled by improving their configuration and using the measuring circuit designs with high resolving power.

8.7 Conclusions

The slip displacement signal detection method considered in the present paper furnishes an explanation of the main detection principles and allows robot sensing systems to obtain broad capabilities. The authors developed a wide variety of SDS schemes and mathematical models for capacitative, magnetic and light-reflecting sensitive elements with improved characteristics (accuracy time response, sensitivity). It is very important that the developed multi-component capacity sensor allows identifying the direction of object slippage based on slip displacement signal detection which can appear in the case of intelligent robot collisions with any obstacle in a dynamic environment. The design of self-clamping grippers is also a very appropriate direction for intelligent robot development.

The results of the research are applicable in the automatic adjustment of the clamping force of robot's gripper and robot motion correction algorithms in real time. The methods introduced by the authors may be also used in random operational conditions, within problems of automatic assembly, sorting, pattern and image recognition in the working zones of robots. The proposed sensors and models can be used for the synthesis of intelligent robot control systems [2, 16] with new features and for solving orientation and control tasks during intelligent robot contacts with obstacles.

References

[1] Ravinder S. Dahiya, Giorgio Metta, Maurizio Valle and Giulio Sandini. Tactile Sensing From Humans to Humanoids, volume 26 of Issue 1, pages 1–20. IEEE Transactions on Robotics, 2010.
[2] A. A. Kargin. Introduction to Intelligent Machines. Book 1: Intelligent Regulators. Nord-Press, DonNU, Donetsk, 2010.

[3] Y. P. Kondratenko. Measurements methods for slip displacement signal registration. In Proc. of Intern. Symposium on Measurement Technology and Intelligent Instruments, pages 1451–1461, Chongqing-Wuhan, China, 1993. Published by SPIE. USA.

[4] Y. P. Kondratenko and X. Y. Huang. Slip displacement sensors of robotic assembly system. In Proc. of 10-th Intern. Conference on Assembly Automation, pages 429–436, Kanazava, Japan, 23–25 Oct 1989. IFS Publications. Kempston, United Kingdom.

[5] Y. P. Kondratenko and I. L. Nazarova. Mathematical model of capacity sensor with conical configuration of sensitive element. In Proceedings of the Donetsk National Technical University, No. 11(186), pages 186–191, Donetsk: DNTU.

[6] Y. P. Kondratenko and E. A. Shvets. Adaptive gripper of intelligent robot. Patent No. 14569, Ukraine, 2006.

[7] Y. P. Kondratenko and O. S. Shyshkin. Experimental studies of the magnetic slip displacement sensors for adaptive robotic systems. In Proceedings of the Odessa Polytechnic University, pages 47–51, Odessa, 2005. Special Issue.

[8] Y. P. Kondratenko and O. S. Shyshkin. Synthesis of regression models of magnetic systems of slip displacement sensors. Radioelectronic and Computer Systems, (No. 6(25)): 210–215, Kharkov, 2007.

[9] Y. P. Kondratenko, A. V. Kuzmichev and Y. Z. Yang. Robot control system using slip displacement signal for algoritm correction. In ROBOT CONTROL (SYROCO91). Selected papers from the 3-rd IFAC/IFIP/IMACS Symposium, pages 463–469. Pergamon Press, Vienna, Austria. Oxford-NewYork-Seoul-Tokyo, 1991.

[10] Y. P. Kondratenko, E. A. Shvets and O. S. Shyshkin. Modern sensor systems of intelligent robots based on the slip displacement signal detection. In Annals of DAAAM for 2007 & Proceedings of the 18th International DAAAM Symposium, pages 381–382, Vienna, Austria, 2007. DAAAM International.

[11] Y. P. Kondratenko, L. P. Klymenko, V. Y. Kondratenko, G. V. Kondratenko and E. A. Shvets. Slip displacement sensors for intelligent robots: Solutions and models. In Proceedings of the 2013 IEEE 7th International Conference on Intelligent Data Acquisition and Advanced Computing Systems, Vol. 2, pages 861–866. IDAACS, 2013.

[12] Y. P. Kondratenko, N. Y. Kondratenko and V. Y. Kondratenko. Intelligent sensor system. Patent No. 52080, Ukraine, 2010.

[13] Y. P. Kondratenko, O. S. Shyshkin and V. Y. Kondratenko. Device for detection of slip displacement signal. Patent No. 79155, Ukraine, 2007.

[14] Y. P. Kondratenko, V. Y. Kondratenko, E. A. Shvets and O. S. Shyshkin. Adaptive gripper devices for robotic systems. In Mechatronics and Robotics (M&R-2007): Proceeding of Intern. Scientific-and-Technological Congress (October 2–5, 2007), pages 99–105. Polytechnical University Press, Saint-Petersburg, 2008.

[15] Y. P. Kondratenko, V. Y. Kondratenko, I. V. Markovsky, S. K. Chernov, E. A. Shvets and O. S. Shyshkin. Adaptive gripper of intelligent robot. Patent No. 26252, Ukraine, 2007.

[16] Y. P. Kondratenko, Y. M. Zaporozhets, G. V. Kondratenko and O. S. Shyshkin. Device for identification and analysis of tactile signals for information-control system of the adaptive robot. Patent No. 40710, Ukraine, 2009.

[17] M. H. Lee. Tactile sensing: New directions, new challenges. Int. J. of Robotics Research. 19(7), Jul 2000, vol. 19 no. 7, pp. 636–643., Jul 2000.

[18] Mark H. Lee and Howard R. Nicholls. Tactile sensing for mechatronics - a state of the art survey, volume 9, pages 1–31. Mechatronics, 1999.

[19] H. B. Muhammad, C. M. Oddo, L. Beccai, C. Recchiuto, C. J. Anthony, M. J. Adams, M. C. Carrozza, D. W. L. Hukins and M. C. L. Ward. Development of a bioinspired MEMS based capacitive tactile sensor for a robotic finger. Sensors and Actuators A-165, pages 221–229, 2011.

[20] Howard R. Nicholls and Mark H. Lee. A survey of Robot Tactile Sensor Technology. The International Journal of Robotic Research, (Vol. 8, No. 3):3–30, June 1989.

[21] E. P. Reidemeister and L. K. Johnson. Capacitive acceleration sensor for vehicle applications. In Sensors and actuators, pages 29–34. SP-1066, 1995.

[22] Johan Tegin and Jan Wikander. Tactile Sensing in Intelligent Robotic Manipulation-A Review. Industrial Robot: An International Journal, (Vol. 32, No. 1):64–70, 2005.

[23] M. I. Tiwana, A. Shashank, S. J. Redmond and N. H. Lovell. Characterization of a capacitive tactile shear sensor for application in robotic and upper limb prostheses. Sensors and actuators, A-165, pages 164–172, 2011.

[24] M. Ueda and K. Iwata. Adaptive grasping operation of an industrial robot. In Proc. of the 3-rd Int. Symp. Ind. Robots, pages 301–310, Zurich, 1973.

[25] M. Ueda, K. Iwata and H. Shingu. Tactile sensors for an industrial robot to detect a slip. In 2-nd Int. Symp. on Industrial Robots, pages 63–76, Chicago, USA, 1972.

[26] Y. M. Zaporozhets. Qualitative analysis of the characteristics of direct permanent magnets in magnetic systems with a gap. Technical electrodynamics, (No. 3):19–24, 1980.

[27] Y. M. Zaporozhets, Y. P. Kondratenko and O. S. Shyshkin. Three-dimensional mathematical model for calculating the magnetic induction in magnetic-sensitive system of slip displacement sensor. Technical electrodynamics, (No. 5):76–79, 2008.

[28] Y. M. Zaporozhets, Y. P. Kondratenko and O. S. Shyshkin. Mathematical model of slip displacement sensor with registration of transversal constituents of magnetic field of sensing element. Technical electrodynamics, (No. 4):67–72, 2012.

9

Distributed Data Acquisition and Control Systems for a Sized Autonomous Vehicle

T. Happek, U. Lang, T. Bockmeier, D. Neubauer and A. Kuznietsov

University of Applied Sciences, Friedberg, Germany
Corresponding author: T. Happek <t.Happek@gmx.net>

Abstract

In this paper, we present an autonomous car with distributed data processing. The car is controlled by a multitude of independent sensors. For lane detection, a camera is used, which detects the lane marks using a Hough transformation. Once the camera detects these, one of them is selected to be followed by the car. This lane is verified by the other sensors of the car. These sensors check the route for obstructions or allow the car to scan a parking space and to park on the roadside if the gap is large enough. The car is built on a scale of 1:10 and shows excellent results on a test track.

Keywords: Edge detection, image processing, microcontrollers, camera.

9.1 Introduction

In modern times, the question of safe traveling becomes more and more important. Most accidents are caused by human failure, so that in many sectors of industry the issue of "autonomous driving" is of increasing interest. An autonomous car will not have problems like being in a bad shape that day or tiredness and will suffer less from reduced visibility due to environmental influences. A car with laser sensors to detect objects on the road, sensors that measure the grip of the road, that calculate speed based on the signals of these sensors and with a fixed reaction time will reduce the number of accidents and related costs.

Advances in Intelligent Robotics and Collaborative Automation, 193–210.

This chapter describes the project of an autonomous vehicle on a scale of 1:10, which was developed based on decentralized signal routing. The objective of the project is to build an autonomous car that is able to drive autonomously on a scaled road, including the detection of stopping lines, finding a parking area and parking autonomously.

The project is divided into three sections. The first section is about the car itself, the platform of the project. This section describes the sensors of the car, the schematic construction and the signal flow of the car.

The second section is about lane detection, the most important part of the vehicle. Utilizing a camera with several image filters, the lane marks of the road can be extracted from the camera image. This section also describes the calculation of the driving lane for the car.

The control of the vehicle is the matter of the third section. The car runs based on a mathematical model of the car, which calculates the speed and the steering angle of the car in real time, based on the driving lane provided by the camera.

The car is tested on a scaled indoor test track.

9.2 The Testing Environment

Since the car is based on a scaled model car, the test track has to be scaled, too. Therefore, the test track has the same dimensions as the scaled road that is used in a competition for cars on a scale of 1:10 that takes place in Germany every year. As you can see in Figure 9.1, the road has fixed lane marks. This is important, because it's about a prototype. On a real road, several

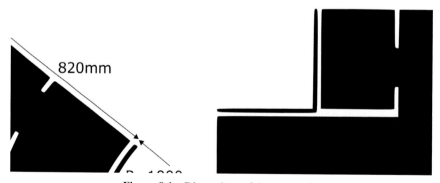

Figure 9.1   Dimensions of the test track.

types of lane marks exist. From white lines, as in the test track, to pillars and a missing center line, every type of lane marks is expected to show up.

In order to simplify the lane detection on the test track, it is assumed, that at all times only one type of lane marks exists. The road has a fixed width and the radius of the curves measure at least 1 meter. The test track has no slope, but a flat surface.

9.3 Description of the System

The basic idea for the vehicle is related to a distributed data acquisition strategy. That means, that all peripherals are not managed by a single microcontroller, but each peripheral has its own microcontroller, which handles the data processing for a specific task. All together are used to analyze the data of the different sensors of the car. Smaller controllers, for instance for the distance sensors or for the camera, are managed by one main controller. The input of the smaller controllers is provided simultaneously via CAN.

The base of the vehicle is a model car scaled 1:10. It includes each mechanical peripheral of a car, like the chassis or the engine. A platform for the control system is added. Figure 9.2 shows the schematic topview of the car with the added platform.

The vehicle itself is equipped with two front boards, a side board, a rear board and the motherboard. All of these boards have one microcontroller for

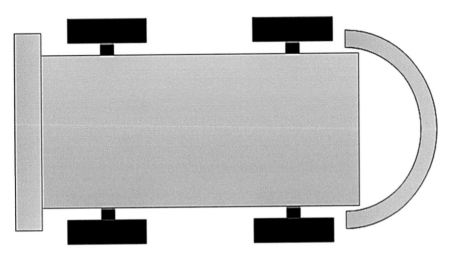

Figure 9.2 Schematic base of the model car.

data analysis that is positioned next to the sensors. The front boards provide infrared sensors for object tracking in the distance.

The infrared sensors of the side board have the task of finding a parking space and transmitting the information via CAN bus to the main controller, which undertakes the control of parking supported by the information it gets from the sensors in the front and back of the car.

The rear board is equipped with infrared sensors, too. It serves the back of the vehicle only. That guarantees a safe distance to all objects in the back. The microcontrollers are responsible for the data processing of each board and send the information to the main controller via CAN bus. Each of the microcontrollers reacts on the incoming input signals of the corresponding sensors according to its implemented control. The infrared sensors are distributed alongside the car as you can see in Figure 9.3.

The motherboard with the integrated main controller is the main access point of the car. It provides the CAN bus connection, the power supply for the other boards and external sensors of the car. But primarily it's the central communications point of the car and manages the information that comes from the peripheral boards, including the data from the external sensors, the control signals for the engine and the servo for the starring angle.

The motherboard gets its power supply from three 5 V batteries. With these three batteries, the model car is able to drive about one hour autonomously.

The main task for the main controller is the control of the vehicle. It calculates the speed of the car and the starring angle based on a mathematical model of the car and the information of the sensors. The external engine driver sets the speed via PWM. The starring angle of the car is adjusted by the front wheels. An additional servo controls the wheel's angle.

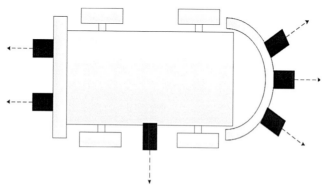

Figure 9.3   The infrared sensors distributed alongside the car.

The camera and its lane detection is the most important component of the vehicle. It is installed in the middle of the front of the car, see Figure 9.4. The viewing angle is important for the position of the camera. If the viewing angle is too small, the pictures of the camera show a near area in front of the car only, but not the area in the middle distance. If the viewing angle is too big, the camera shows a big area in front of the car indicating near and far distances, but the information of the road is so condensed, that an efficient lane detection isn't possible. The angle depends also on the height of the camera and the numerical aperture of the lens. The higher the camera is positioned, the smaller the viewing angle. For this project, the camera has a height of 30 cm and a viewing angle of 35 degrees. The height and the angle of the camera are based on experimental research.

Figure 9.5 shows the reduced signal flow of the vehicle. The information from the infrared sensors is sent to a small microcontroller, as it is visualized by the spotted lines. In reality, each sensor has its own microcontroller, but to reduce the complexity of the graphic, they were shown as one. The camera has its own microcontroller. This controller must be able to accomplish the necessary calculations for lane detection in time. For the control of the vehicle by the main controller, it is necessary that all information from all other controllers are actualized in one calculation step, this is needed for the mathematical model of the car. The main controller gathers the analyzed data provided by the smaller microcontrollers, the data from the camera

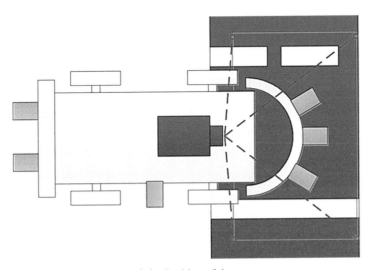

Figure 9.4   Position of the camera.

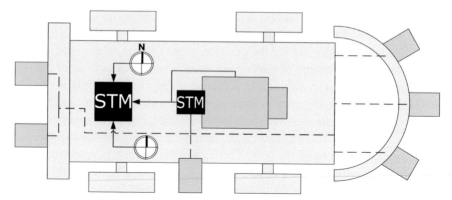

Figure 9.5   Schematic signal flow of the vehicle.

about the driving lane and the information from other sensors like gyroscope and accelerometer for its calculation. The essential signal flow of all these components to the main controller is visualized by the solid lines in Figure 9.5. After its calculation, the main controller sends control signals to the engine and the servo, which controls the starring angle of the car.

Incremental encoders on the rear wheels detect the actual speed and calculate the path the vehicle has traveled during the last calculation step of the mathematical model. The sensors send the data via CAN bus to the main controller. The vehicle is front-engined, so traction of the rear wheels is ensured. Potential error in measurement through spinning is avoided.

There are two modules that do not communicate via CAN bus with the main controller: the first one is the camera, ensuring that the vehicle keeps the track, the second is a sensor module, which includes the gyroscope and accelerometer. Both modules do not have a CAN interface, but they communicate via an USART interface with the main microcontroller.

In the future, the focus will be on an interactive network of several independent vehicles based on radio transmission. This will allow all vehicles to communicate with each other and share information like traction and behavior of the road, actual position from GPS, or speed. The radio transmission is carried out with the industry standard called "Zigbee". An XBEE module of the company "Digi" undertakes the radio transmission. The module uses an UART interface for the communication with the main microcontroller on the vehicle. Via this interface, the car will get information from other cars nearby. A detailed overview of the data processing system, including the XBEE module, is shown in Figure 9.6.

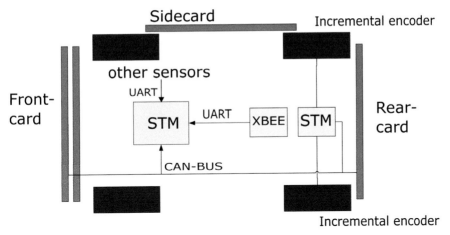

Figure 9.6 Overview of the data processing system.

9.4 Lane Detection

There are several steps needed to accomplish the lane detection.

First, the image has to be analyzed with an In-Range filter. In the second step, the points that the Hough-transformation has identified as lane marks, are divided into left and right lane marks. Next, the least squares method is used to transform the lane marks into a second-degree polynomial, thus providing the base to calculate the driving lane. Subsequently, the points of the driving lane are transformed into world coordinates.

Two types of filters are used to get the needed information from the image. Both are functions from the OpenCV-library. An In-Range filter is used to detect the white lane marks on the defined test track. The Hough-transformation calculates the exact position of the lane marks preparing them for the next steps.

9.4.1 In-Range Filter

The In-Range filter transforms an RGB-image into an 8-bit binary image. It's made for the detection of pixels in a variable color range. The transformed picture has the same resolution as the original picture. Pixels belonging to the chosen color range are white. All other pixels in the image are black. The function works with the individual values of the RGB format. The chosen color is defined by two critical values of this format.

Figure 9.7 shows the result of the In-Range filter.

Figure 9.7   Comparison between original and in-range image.

9.4.2 Hough-Transformation

The Hough-transformation is an algorithm to detect lines or circles in images, which in this case means that it investigates the binary image from the In-Range filter in order to find the lane marks.

The Hessian normal form converts individual pixels, so that they can be recognized as lines in the Hough space. In this state, space lines are expressed by the distance to the point of origin and the angle to one of the axes. Due to the fact that the exact angle of the marks is unknown, the distance to the point of origin is calculated based on Equation (9.1), utilizing the most probable angles:

$$r = x \cdot \cos(a) + y \cdot \sin(a). \tag{9.1}$$

The intersection of the sinusoidals provides an angle and the distance of the straight line from the origin of coordinates. These parameters create a new line, so that the majority of the pixels can be detected. Furthermore, the function from the OpenCV-library returns the start and the endpoint of each Hough-line. As Figure 9.8 shows, the lines of the Hough-transformation are precisely mapped on the lane marks of the road.

Figure 9.8   Original image without and with Hough-lines.

9.4.3 Lane Marks

To provide a more precise calculation, all points along the line are included. These points are stored in two arrays and then sorted. As a first sorting criterion, the position of the last driving lane is used. The second criterion for sorting derives from their position in the image.

As mentioned before, the information in the image regarding long distances can be critical depending on the viewing angle and height of the camera. In order to concentrate on noncritical information only, points in the middle area of the image are used. Figure 9.9 shows the sorted points on the right and the corresponding Hough-lines on the left side.

9.4.4 Polynomial

To describe the lane marks more efficiently, a second-degree polynomial is used. The coefficients of the parable are derived by the least-squares method. A polynomial of a higher degree isn't needed, because the effort to calculate the coefficients is too high to make sense in this context, for the speed of the image processing is one of the critical points of the project. Furthermore, the area of the road, which is pictured by the camera, is too small. The road is unable to clone the typical form of a third-degree polynomial.

As visible in Figure 9.10, the parables derived from the sorted points are mapped precisely on the lane marks of the road. The algorithm to calculate the coefficients derived from the points of the lane marks is handwritten.

9.4.5 Driving Lane

The driving lane for the car lies between the parables mentioned in the last chapter. To calculate the position of the points of the driving lane, the average

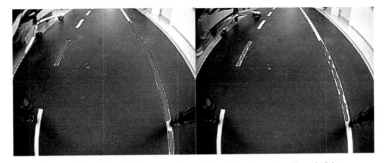

Figure 9.9   Hough-Lines and sorted points along the Hough-Lines.

Figure 9.10   Sorted Points and Least-Square Parable.

of two opponent points of the two parables is taken. According to 9.2, the average for the x- and y-coordinates is calculated.

$$\begin{pmatrix} x_m \\ y_m \end{pmatrix} = \begin{pmatrix} x_1 \\ y_1 \end{pmatrix} + \frac{1}{2} \left(\begin{pmatrix} x_2 \\ y_2 \end{pmatrix} - \begin{pmatrix} x_1 \\ y_1 \end{pmatrix} \right). \qquad (9.2)$$

In order to simplify the transformation from pixel-coordinates to world-coordinates, the driving lane is described by a fixed number of points in the image. The essential feature of these points is that they lie in predefined rows in the image. So, there is only the need to calculate the horizontal position of the parable for these points.

Theoretically it is possible that the program delivers an incorrect driving lane. Mistakes can occur because of flash lights, reflections on the road, missing lane marks due to different reasons or extreme light conditions, which are much faster than the auto white balance of the camera can bear. So in order to avoid mistakes that occur within a short time period, some kind of stabilization is required. Short time in this case means shorter than one second.

For the purpose of stabilization, the different driving points are stored. The stabilization works with these stored points in combination with four defined edge points in the image. First, the algorithm checks if the edge points of the new image differ from the edge points in the old image.

If the difference between the old points and the new points is low, the driving lane is calculated and the driving points are stored. In case that the difference between the points is too big, the driving lane is not updated and the driving lane is calculated by using the stored points. The algorithm works with the changes of the stored points. The new points are calculated by using the difference between the last image and the current one. This difference is derived from the change of the difference between the third and second image, that have been taken before the current one, and the difference between the second and the first image before the current one.

The critical values for the difference also depend on this calculation. That means that in curvas, the critical values are higher. If not, only the last three images are used for the calculation, in order to reduce the noise of the driving lane. However, in this case, the reaction time of the algorithm is lesser.

The reaction time also depends on the fps (frames per second) of the camera. For this project, a camera with 100 fps is used and the last fifteen driving lanes are stored. The number of stored driving lanes for 100 fps is based on experimental research.

Figure 9.10 shows the driving lane in red color. The four edge points mark the edge points of the rectangle.

9.4.6 Stop Line

One of the main tasks of the camera is to detect stop lines. Figure 9.11 shows the dimensions of the stop lines for this test track.

In order to detect stop lines, the algorithm is searching for the main characteristics. First, the stop line is a horizontal line in the image. If the angle of a vertical line in the image is defined as zero degrees, that means, that the perfect stop line has an angle of 90 degree. The algorithm not only searches for 90 degree lines. The angle for a potential stop line is smaller than −75 degree and bigger than +75 degree.

The next criterion of a stop line is that it lies on the car traffic lane. So, the algorithm does not need to search in the complete image for stop lines, but only in the area of the cars traffic lane. This area is marked by the four edge points of the rectangle mentioned in the last chapter. Once the algorithm

Figure 9.11   Parables and driving lane.

finds a potential stop line in the right area with a correct angle, the algorithm checks the next two characteristics of a stop line: the length and the width of the stop line.

The length of the stop line is easy to check. The stop line must be as long as the road is wide, so the algorithm only needs to check the endpoints of the line. On the left side, the endpoint of the stop line must lie on the middle road marking. On the right side, the stop line borders on the left road marking from the crossing road. The stop line and the road marking differ in just one point: the width.

Since it is not possible to perceive the differences of the width in each situation, the stop line has no defined end point on this side. So, the algorithm checks if the end point of the potential stop line lies on or above the right road marking. It is hard to measure the width of a line in an image that has constant width and length in reality. The width of the line in the image in pixels depends on the camera position in relation to the line, the numerical aperture of the camera lens and the resolution of the camera. So, because in this project the position of the camera changes from time to time, measuring the width is not reliable to perceive the stop line. Therefore, the width is not used as a criterion for stop lines.

Figure 9.12 shows a typical crossing situation. The left image visualizes the basic situation and the middle image shows the search area as a rectangle. Here you can see that the stop line on the left side is not covered by the research area so the algorithm doesn't recognize the line as a stop line. On the right image, the stop line ends correctly on the middle road marking. The line in the image shows that the algorithm has found a stop line. Due to the left road marking from the crossing road, the line ends outside the real stop line.

9.4.7 Coordinate Transformation

To control the car, the lateral deviation and the course angle are needed. Both are calculated by the controller of the camera. The scale unit for the

Figure 9.12   Detection of stop lines.

lateral deviation is meters and degrees for the course angle. Course angle means the angle of the driving lane which is calculated by the camera. The lateral deviation is the distance of the car's center of gravity to the driving lane when they are at the same level. Since the lateral deviation is needed in meters, the algorithm has to convert the pixel coordinates from the image into meters in the real world. The course angle can be calculated from the pixel coordinates in the image, but this method is error-prone.

There are two different methods to convert the pixels into meters.

Pixels can be converted via Equations (9.3) and (9.4).

$$x(u, v) = \frac{h}{\tan\left[\left(\bar{\theta} - \alpha\right) + u\frac{2\alpha}{n-1}\right]} \cdot \cos\left[\left(\bar{\gamma} - \alpha\right) + u\frac{2\alpha}{n-1}\right] + l, \quad (9.3)$$

$$x(u, v) = \frac{h}{\tan\left[\left(\bar{\theta} - \alpha\right) + u\frac{2\alpha}{n-1}\right]} \cdot \cos\left[\left(\bar{\gamma} - \alpha\right) + u\frac{2\alpha}{n-1}\right] + l. \quad (9.4)$$

In the equations, x and y are the coordinates in meters. $\bar{\gamma}$ stands for the drift angle of the camera in the plane area and $\bar{\theta}$ stands for the pitch angle of the camera. α is the numerical aperture of the camera, u and v are the coordinates of one pixel in the image.

Using this equation, the complete image can be converted into real-world coordinates. The drawback of this method is that all parameters of the camera have to be known exactly; every difference between the numerical aperture in the equation and the exact physical aperture of the camera lens can cause massive failure in the calculation. Furthermore, this method needs more calculation time on the target hardware. A big plus of this method is that the camera can be re-positioned during experimental research.

The second method is to store references to some pixels in lookup tables. For these pixels, the corresponding values in meters can be calculated or can be measured. This method expends much less calculation time but is also much less precise. With this method, the camera cannot be re-positioned during experiment research. Every time the camera is re-positioned the reference tables must be re-calculated.

The method to prefer depends on the project requirements regarding accuracy and the projects hardware. For this project, the second method is used. To meet the demands on accuracy, for each tenth pixel of the camera, a reference is stored.

9.5 Control of the Vehicle

The driving dynamic of the vehicle is characterized by the linear track model of Ackermann. As Figure 9.13(a) shows, the model is composed of a rear and front wheel which are connected by an axe. In order to rotate the vehicle on its main axe, the steering angle can be set with the front wheel.

To reduce the complexity of vehicle dynamics, three simplifications are made.

These are:

- Neglect the air resistance, because the vehicle speed is very low;
- Lateral forces on the wheels are linearized;
- No roll of the vehicle about the x and y axis.

Using these simplifications, the created model should differ only marginally from reality. Linking the transverse dynamics of the vehicle with the driving dynamics, you can derive the following relation:

$$
r \begin{bmatrix} \ddot{\psi}(t) \\ \dot{\theta}_\Delta(t) \\ \dot{\gamma}(t) \end{bmatrix} = \begin{bmatrix} a_{22} & 0 & 0 \\ -1 & 0 & 0 \\ 0 & V & 0 \end{bmatrix} \cdot \begin{bmatrix} \dot{\psi}(t) \\ \theta_\Delta(t) \\ \gamma(t) \end{bmatrix} + \begin{bmatrix} b_2 \\ 0 \\ 0 \end{bmatrix} \cdot \delta(t). \qquad (9.5)
$$

Because of the equation in the state space, a controller can be designed using tools such as Matlab Simulink or Scilab X-cos.

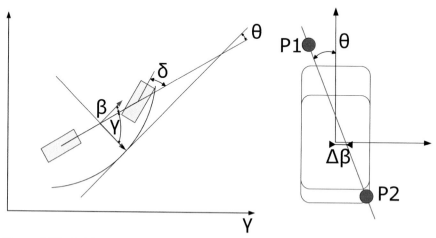

Figure 9.13   Driving along a set path: Track model (a); Lateral deviation and heading angle (b).

In order to keep the vehicle on the track, the conditions such as heading angle, the yaw rate and lateral deviation must be known. A gyroscope is used to detect the yaw rate of the vehicle. The lateral deviation and the course angle are calculated from the camera. The camera sends the lateral deviation and the curse angle. Until the next image is analyzed, the coordinates on the microcontroller stay the same as before. Between two pictures the lock angle and the transverse deviation were recalculated after each motion. This is possible because the velocity and yaw rate are known at any time. Figure 9.13(b) illustrates the relationship of lateral deviation (ΔY) and heading angle (α).

9.6 Results

This section gives an overview of the project results.

The autonomous car was built with the hardware suggested before. Experiments on scaled test roads show that the car can drive autonomously. However, the tests also showed the limitations of this prototype. The effort for the image processing was undervalued. The on-board processor of the camera isn't able to accomplish the necessary calculations in time. In terms of reaction to this fact, the maximum speed of the car has to be very slow. If it isn't, the control of the vehicle gets unstable with a more or less random driving path. In addition, the car has problems with too sharp curves. The process to divide the image isn't dynamic, so in curves the preset section becomes incorrect and the algorithm isn't able to calculate a correct driving path. Thanks to its laser sensors, the car is able to avoid collisions with baffles.

To improve the performance of the car, the hardware for the image processing has to be improved. The image processing works stable. Problems derive from the calculation algorithm of the driving path. At this point in time, the algorithm doesn't contain the necessary interrupts for every situation on the road, but this drawback will be corrected in the second prototype.

9.7 Conclusions

In this chapter, an autonomous vehicle with distributed data acquisition and control systems has been presented. For control, the vehicle has a number of independent sensors. The main sensor is a camera with a lane tracking algorithm, which contains edge detection and Hough transformation. The lane is verified by laser sensors in the front and side of the vehicle. It is planned to build a superordinate control system, which leads a group of autonomous vehicles using a wireless communication protocol.

References

[1] J. Canny, 'A Computational Approach to Edge Detection', IEEE 1986.

[2] A. Erhardt, 'Einführung in die Digitale Bildverarbeitung', Offenburg 2008.

[3] W. Heiden, 'Kanten in Bildern - Filterung und Kantenerkennung', St. Augustin 2009.

[4] B. Jähne, 'Digitale Bildverarbeitung und Bildgewinnung', Berlin 2013.

[5] M. Jauernig, 'Einsatz von Algorithmen der Photogrammmetrie und Bildverarbeitung zur Einblendung spezifischer Lichtraumprofile in Videosequenzen', Hochschule Leipzig 2006.

[6] A. Kant, 'Bildverarbeitungsmodul zur Fahrspurerkennung für ein autonomes Fahrzeug', Hochschule Hamburg 2007.

[7] L. F. Kirk, 'Schätzung der Brennweite mit Hilfe der Hough-Transformation', Fachhochschule, Köln 2011.

[8] N. Kruse, 'Kameragestützte Fahrspurerkennung für autonome Modell-fahrzeuge', Universität Hamburg 2008.

[9] G. Linß, 'Praktische Ausbildung und Training Qualitätsmanagement Objekterkennung mit Hough-Transformation', Technische Universität Ilmenau.

[10] R. Maini, H. Aggarwal, 'Study and Comparison of Various Image Edge Detection Techniques', Punjabi University, India.

[11] P. Schöley, Kantendetektoren, Technische Universität Dresden, 2011.

[12] J. Unger, 'Untersuchung von Linien und Kantenextraktionsalgorith-men im Rahmen der Verifikation von Ackerlandobjekten', Universität Hannover, 2009.

[13] C. Wagner, Kantenextraktion – Klassische Verfahren, Universität, Ulm, 2006.

[14] G. M. Wagner, G. M., 'Bestimmung der Kameraverzerrung mit Hilfe der Hough-Transformation', Fachhochschule Köln, 2011.

[15] L. Bergen, H. Burkhardt, Morphological image processing.

[16] J. Wohlfeil, 'Detection and tracking of vehicles with a moving camera' Humbolt-Universität zu Berlin Institut für Informatik; Deutsches Zentrum für Luft- und Raumfahrt Institut für Verkehrs-forschung, 2006.

[17] B. Hulin 'Video-based obstacle detection clearance of the pantograph electric railways', Fakultät für Elektrotechnik und Informationstechnick, Technische Universität München, 2003.

[18] J. Alberts, 'Vision for the project FAUST, Focus of Expansion/optical flow', Hamburg University of Applied Sciences, 2007.

[19] Tutorial Filter; Vision&Control System Lighting Optics, 2007.

[20] Prof. Dr. –Ing. Habil G. Linß, 'Practical education and training quality management, object detection with Hough transform', Technische Universität Ilmenau, Department of Mechanical Engineering, Department of Quality Assurance.

[21] C. Wagner, Edge extraction, Classical methods; Seminar presentation on "Image segmentation and computer Vision", 2006.

[22] I. Nikolov, 'Adaptive camera parameters for optimum lane detection and tracking', Hamburg University of Applied Sciences, 2007.

[23] V. Schaefer, A. Zipser, 'Algorithmic Applications Hough transform', 2006.

[24] C.Rathemacher, 'GPU-based detection of algebraic structures writable', Fachhochschule Wiesbaden, Fachbereich Design Informatik Medien, 2007.

[25] M. Pelkofer, 'Behavioral decision for autonomous vehicles with gaze control', Universität der Bundeswehr München, Fakultät für Luft- und Raumfahrttechnik, Institut für Systemdynamik und Flugmechanik, 2003.

[26] D.Berger, 'Lane inference system for lane algorithm Three Feature Based Lane Detection Algorithm (TFALDA)', Hamburg University of Applied Sciences, 2008.

[27] Alberto Broggi, 'A Massively Parallel Approach to Real-Time Vision-Based Road Markings Detection', Dipartimento di ingengneria dell'Information Universita di Parma.

[28] Andreas Weide, 'Entwicklung einer Kameragestützten Fahrspurerkennung für ein autonomes Fahrzeug', University of Applied Sciences (Friedberg, Germany), 2013.

10

Polymetric Sensing in Intelligent Systems

Yu. Zhukov, B. Gordeev, A. Zivenko and A. Nakonechniy

National University of Shipbuilding named after admiral Makarov, Ukraine
Corresponding author: Yu. Zhukov <prof.zhukov@gmail.com>

Abstract

The authors examine the up-to-date relationship between the theory of polymetric measurements and the state of the art in intelligent system sensing.
The chapter contains commentaries about: concepts and axioms of polymetric
measurements theory, corresponding monitoring information systems used in
different technologies and some prospects for polymetric sensing in intelligent
systems and robots. The application of the described concepts in technological
processes ready to be controlled by intelligent systems is illustrated.

Keywords: Polymetric signal, polymetric sensor, intelligent sensory system,
control, axiomatic theory, modelling.

10.1 Topicality of Polymetric Sensing

For some time it has been widely recognized [1] that the future factory should
be a smart facility where design and manufacture are strongly integrated into a
single engineering process that enables 'right first time' every time production
of products to take place. It seems completely natural to use intelligent robots
and/or control systems at such factories to provide on-line, opportune and
precise assessment of the quality of the production process and to guarantee
the match and fit, performance and functionality of every component of the
product prototype that is created.

Wide deployment of sensor-based intelligent robots at the heart of future
products and technology-driven systems may substantially accelerate the

Advances in Intelligent Robotics and Collaborative Automation, 211–234.

process of their technological convergence and lead to their introduction in similar processes and products.

At the same time it is well known [2] that intelligent robot creation and development is a rather fragmented task and there are different progress-restricting problems in each particular field of research. Numerous advanced research teams are doing their best to overcome these limitations for intelligent robot language, learning, reasoning, perception, planning and control.

Nowadays, a few researchers are still developing the Ideal Rational Robot, as it is quite evident that the computations necessary to reach ideal rationality in real operating environments require much more time and much more productive processors. Productivity is permanently growing but being still limited for IRR tasks. Furthermore, the robotic perception of all the necessary characteristics of real operating environments has not been sufficiently provided yet.

The number of corresponding sensors and instruments necessary for these robots is growing, but their sensing remains too complicated, expensive, time-consuming and IRR remains far from being realistic for actual application.

F.H. Simons' concept of Bounded Rationality and the concept of Asymptotic Bounded Optimality based on S.J. Russell's and D. Subramanian's approach of provably bounded-optimal agents have formed a rather pragmatic and fruitful trend for the development of optimal programs.

One has to take into account the fact that time limitations are even stricter if we are trying to overcome them for both calculation and perception tasks. While immense progress has been made in each of these subfields in the last few decades, it is still necessary to comprehend how they can be integrated to produce a really effective intelligent robot.

The introduction of the concepts of agent-based methods and systems [3, 4] including holonic environment [5] and multi-agent perceptive agencies [6] jointly with the concept of polymetric measurements [7–9] has engendered a strong incentive to evaluate all the potentialities of using polymetric sensing for intelligent robots, as this may be a precondition for generating real benefits in the field.

The matter is that obvious practical success has been achieved during the wide deployment of SADCO® polymetric systems for monitoring a variety of multiple quantitative and qualitative characteristics of different liquid and/or loose cargoes using a single polymetric sensor for measuring more than three characteristics of cargoes simultaneously within a single chronostopos framework. Similar information systems were also a successful tool for the online remote control of complex technological processes of production, storage and consumption of various technological media [8].

But, as indicated above, there are very specific requirements for sensing in intelligent robots. In fact, one of the most important restrictions is connected with very limited time-consumption for the input of real-time information concerning an intelligent robot and/or multi-agent control systems operational environment.

Thus, we face an actual and urgent need to integrate and combine these advanced approaches within the calculation and perception components of intelligent robots and/or multi-agent monitoring and control system design, starting from different (nearly diametrically opposite) initial backgrounds of each component and, by means of irrefutable arguments, arriving at jointly acceptable conclusions and effective solutions.

10.2 Advanced Perception Components of Intelligent Systems or Robots

10.2.1 Comparison of the Basics of Classical and Polymetric Sensing

Classical or industrial metrology is based on the contemporary Axiomatic Theory [10, 11].

In classical industrial metrology, it is presupposed that for practical measurement processes it is necessary to have some set $\Omega = \{\sigma, \lambda, \ldots\}$ of different instruments with various sensor transformation and construction systems.

But every particular instrument has a custom-made uniform scale for the assessment of the actual value of the object specific characteristic under control. Let $a_\sigma(i)$ be a numerical function of two nonnumeric variables – a physical object $i \in \aleph$ and a specific instrument, i.e.

$$a : \Omega \times \aleph \to \Re \ and \ (\sigma, i) \mapsto a_\sigma(i). \tag{10.1}$$

This function is called a quantitative assessment of a physical quantity of an object.

In the general case for an industrial control system or for intelligent robot sensing process, it is necessary to have N instruments for each characteristic under monitoring at M possible locations of the components of the object under control. The more practical the application, the quicker we face the curse of multidimensionality for our control system.

Polymetric measurement in general is the process of getting simultaneous assessments of a set of object physical quantities (more than two) using one special measuring transformer (a sensor). The first successful appraisal of the corresponding prototype instruments was carried out in 1988–1992 on-board three different vessels during full-scale sea trials. After the successful tests and the recognition of the instruments by customers and the classification societies, the theoretical background was generalized and presented to the scientific community [12–14].

The latest results [7–9] seem to be prospective for intelligent robot sensing due to reduced time and financial pressure, simplified design and reduced general number of sensory components.

Summarizing the comparison of the basics of classical and polymetric measurement theories, it is essential to comment another consequence of their axioms and definitions. The introduction of the principle of the simultaneous assessment of a physical quantity and its measurement from the same polymetric signal Polymetric signal is one of the key provisions of the theory and the design practice for developing appropriate instruments. The structure of an appropriate perceptive intelligent control or monitoring system should be changed correspondingly.

10.2.2 Advanced Structure of Multi-Agent Intelligent Systems

In order for multi-agent control or monitoring systems and intelligent robots to satisfactorily fulfil the potential missions and applications envisioned for them, it is necessary to incorporate as many recent advances in the above described fields as possible within the real-time operation of the intelligent system or robot-controlled processes.

This is the challenge for the intelligent robotics engineer, because many advanced algorithms in this field still require too much time for computation, despite improvements made in recent years in microelectronics and algorithms. Especially, it concerns the problem of intelligent robot sensing (timely and precise perception).

The well-known variability versus space (topos) and versus time (chronos) is related to similar variability of the measured data [6, 7, 10] engendering the advantages of using polymetric measurements.

That is why, in contrast to the multi-sensor perceptive agency concept [6] based on the use of several measuring transformers, each one of them being the sensing part of each particular agent within the distributed multi-sensor control system (i.e. several perceptive agents in complex perceptive agency), the use of one equivalent polymetric transformer for an equivalent perceptive agency is proposed.

The concept of Polymetric Perceptive Agency (PPA) for intelligent system and robot sensing is schematically illustrated in Figure 10.1. Such simplified structure of PPA sub-agency of the Decision-making Agency (DMA) is designated to be used in different industries and technologies (maritime, on-shore, robotics, etc. [9, 15]).

There are some practical optimistic examples of the successful deployment and long-term (more than 15 years) operation of industrial polymetric monitoring and control systems/agencies based on polymetric measurement-Polymetric measurement technique in different fields of manufacturing and transportation [8].

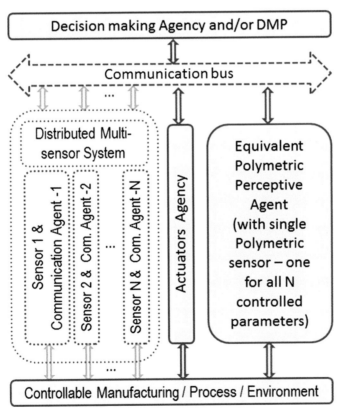

Figure 10.1 The main idea of the replacement of the distributed multi-Sensor system by a polymetric perceptive agent.

10.3 Practical Example of Polymetric Sensing

Here we describe a practical case from the research of the Polymetric Systems Laboratory at the National University of Shipbuilding and LLC AMICO (Mykolaiv, Ukraine). The practical goal is to ensure the effective and safe control of the water level in the cooling pond of nuclear power stations at the spent nuclear fuel storage during normal operation and emergency post-accident operation. There are many level sensors, which are used by the control systems in the normal operation mode: floating-type, hydrostatic, capacitive, radar, ultrasonic, etc. [16]. But there exist many problems concerning their installation and functioning under real post-accident conditions. The matter is that high pressure and extremely high temperature, saturated steam and radiation, vibration and other disturbing factors are expected in the cooling pond in emergency mode.

Thus, high reliability and radiation resistance are the most important requirements for such level-sensing equipment. One of the most suitable sensing techniques in this case is the proposed modified Time Domain Reflectometry (TDRTdr) – see Figure 10.2.

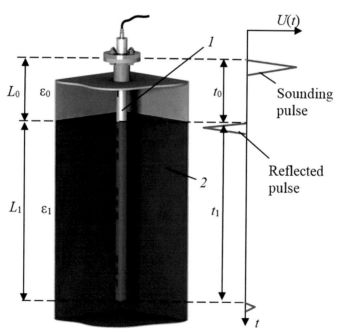

Figure 10.2   TDR Coaxial probe immersed into the liquid and the corresponding polymetric signal.

In this type of level measurement, microwave pulses are conducted along a cable or rod probe *1* partially immersed in the liquid *2* and reflected by the product surface.

Sounding and reflected pulses are detected by the transducer. The transit time of the pulse t_0 is a function of the distance from the level sensor to the surface of the liquid L_0. This time is measured and then the distance from the level sensor to the surface of the liquid is calculated according to the calibration function:

$$L_0 = f(t_0), \tag{10.2}$$

which is usually presented as a linear function:

$$L_0 = b_0 t_0 + b_1, \tag{10.3}$$

where b_0 and b_1 – coefficients which are obtained during a calibration procedure.

From the physical point of view, the slope of this function stands for the speed of the electromagnetic wave which propagates forward from the generator on the pcb-board, along the cable and the specially designed measuring probe, and back to the generator. It can be defined as:

$$b_0 = \frac{c}{2\sqrt{\varepsilon_0}}, \tag{10.4}$$

where c – speed of light in vacuum; ε_0 – dielectric constant of the material through which the electromagnetic pulse propagates (it is close to 1 for air); the coefficient of ½ stands for the fact that the electromagnetic pulse propagates along double the length of the probe (forward and backward).

The fact of the presence and the value of the intercept b_1 are caused by many reasons. One of them is that the electromagnetic pulse actually passes the distance greater than the measuring probe length. In the general case, the real calibration function is not linear [8].

10.3.1 Adding the Time Scale

In this practical case, the basic approaches for constructing the polymetric signal are described based on the TDR technique. The concept of forming an informative pulse polymetric signal includes the following: generation of short pulses and sending them to a special measuring probe, receiving the reflected pulses and signal pre-processing for its final interpretation. It is worth mentioning that in terms of polymetrics, each of these steps is specially designed to increase the informativity and interpretability of the resulting signal.

Therefore, to calculate the distance from the sensor to the surface of the liquid, it is necessary to carry out the calibration procedure. The level-sensing procedure can be simplified by adding additional information to the initial «classic» signal to obtain the time scale of this signal. A stroboscopic transformation of the real signal is one of the necessary signal pre-processing stages during the level measurement procedure.

This transformation is required to produce an expanded time sampled signal for its future conversion to digital form. This transformation can be carried out with the help of a stroboscopic transformer based on two oscillators with frequencies f_1 (the frequency of input signal) and f_2 (the local oscillator frequency) that are offset by a small value $\Delta f = f_1 - f_2$ [17]. The duration of the input signal is:

$$T_1 = \frac{1}{f_1}. \tag{10.5}$$

As a result of this transformation, we have expanded signal duration:

$$T_{TS} = \frac{1}{\Delta f} = \frac{1}{f_1 - f_2}. \tag{10.6}$$

In this case, the relationship between the duration of transformed and the original signal is expressed by the stroboscopic transformation ratio:

$$K_{TS} = \frac{T_{TS}}{T_1} = \frac{f_1}{f_1 - f_2}. \tag{10.7}$$

The next processing step of the transformed signal is the conversion of this inherently analog signal to digital form with the help of the analog-to-digital converter (ADC).

The time scale and delays during analog-to-digital conversion are known. Therefore, it is possible to count ADC conversion cycles and to calculate the time scale of the converted signal. It is not convenient because conversion cycle duration is connected with the ADC parameters, clock frequency value and stability, etc. In order to exclude the use of the conversion cycles number and ADC parameters, it is possible to «add» the additional information about the time scale of the converted signal.

In this case, we can add a special marker which helps to measure the reference time interval – see Figure 10.3.

The main idea is that ADC conversion time T_{ADC} must be greater than the duration of the signal from the output of the stroboscopic transformer T_{TS} to obtain at least 2 sounding pulses in the resulting digitized signal. It is necessary

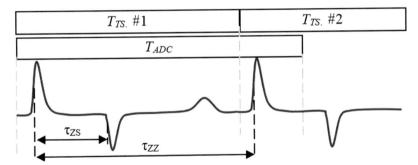

Figure 10.3 Time diagrams for polymetric signal formation using additional reference time intervals.

to count the delays between two sounding pulses τ_{ZZ} and between the first sounding pulse and the reflected pulse τ_{ZS} (in terms of the number of ADC readings).

The delay τ_{ZZ}, expressed in ADC readings count, corresponds to the time delay T_{TS} (in seconds). The next equation should be used to calculate the time scale of the transformed and digitized signal:

$$K_{DT} = \frac{T_{TS}}{\tau_{ZZ}} s/\text{reading}. \tag{10.8}$$

It is possible to calculate the time value of the delay between the pulses t_0 (see Figure 10.2):

$$t_0 = \frac{K_{DT}\tau_{ZS}}{K_{TS}} = \frac{\tau_{ZS}}{f_1\tau_{ZZ}} s. \tag{10.9}$$

Finally, to calculate the distance to the surface of the liquid, it is necessary to use the equation:

$$L_0 = \frac{c}{2\sqrt{\varepsilon_0}} \frac{\tau_{ZS}}{f_1\tau_{ZZ}} + b_1, \tag{10.10}$$

where b_1 (zero shift) is calculated using information on the PCB layout and generator parameters or during a simple calibration procedure.

10.3.2 Adding the Information about the Velocity of the Electromagnetic Wave

In the previous paragraph, a special marker was added to the signal to obtain the time scale of the signal and to easily calculate the distance to the surface of the liquid in normal operating mode.

But, as it was mentioned above, the emergency operating mode can be characterized by high temperatures and pressure, the presence of saturated steam in the cooling pond. Steam under high pressure and temperature will slow down the propagation speed of a radar signal which can cause an additional measurement error. It is possible to add the additional information to the signal for automated correction of measurement. A special reflector or several reflectors are to be used in this case [8]. Almost any step of wave impedance discontinuities can be used as a required reflector (probes with stepwise change of the impedance, fixing elements, etc.). The reflector is placed at a fixed and known distance LR from the generator and the receiver of short electromagnetic pulses GR in the vapour dielectric – presented in Figure 10.4.

If there is vapour in the tank, the pulses reflected from the special reflector and from the surface of the liquid are shifted according to the change of the dielectric constant of vapour (as compared to the situation when there is no vapour in the tank).

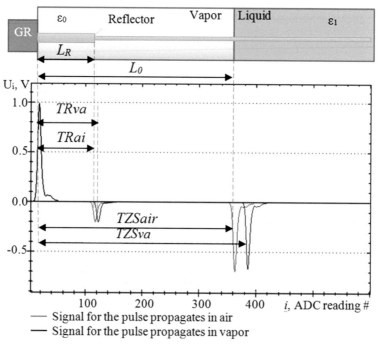

Figure 10.4 Disposition of the measuring probe in the tank, position of the reflector and corresponding signals for the cases with air and vapour.

The delays between the sounding pulse and the pulse reflected from the special reflector τ_R and between the sounding pulse and the pulse reflected from the surface of the liquid τ_{ZS} for cases with air and vapour are different.

The dielectric constant ε_0 can be calculated using the known distance L_R. Therefore, the distance to the surface of the liquid L_0 can be calculated using the corrected dielectric constant value:

$$L_0 = \frac{(L_R - b_1)\tau_{ZS}}{\tau_{ZR}} + b_1. \qquad (10.11)$$

As it can be seen from the equation, the result of the measurement depends on the time intervals between the sounding and reflected pulses and the reference distance between the generator and the reflector.

The above-described example of polymetric signal formation showed that a single signal carries information about the time scale of this signal, the velocity of propagation of electromagnetic waves in dielectrics and the linear dimensions of these layers. This list of measurable parameters can be easily continued by using additional information in the existing signals and building new «hyper-signals» [9] for measuring some other characteristics of controllable objects (e.g. on the basis of the spectral analysis of these signals the controllable liquid can be classified and some quality characteristics of the liquid can be calculated).

10.4 Efficiency of Industrial Polymetric Systems

10.4.1 Naval Application

One of the first polymetric sensory systems was designated for on-board loading and safety control (LASCOS) of tankers, fishing, offshore, supply and research vessels. The prototypes of these systems were developed, industrialized and tested during full-scale sea trials in the early 90-es of the last century [18]. These systems have more developed polymetric sensing subsystems (from the "*topos-chronos*" compatibility and accuracy points of view). That is why they are also successfully providing commercial control of cargo handling operations.

The structure of the hardware part of LASCOS for a typical offshore supply vessel is presented in Figure 10.5. It consists of the following: operator workplace (1); a radar antenna (2); an on-board anemometer (3); the radar display and a keyboard (4); a set of sensors for ship draft monitoring (5); a set of polymetric sensors for fuel-oil, ballast water and other liquid

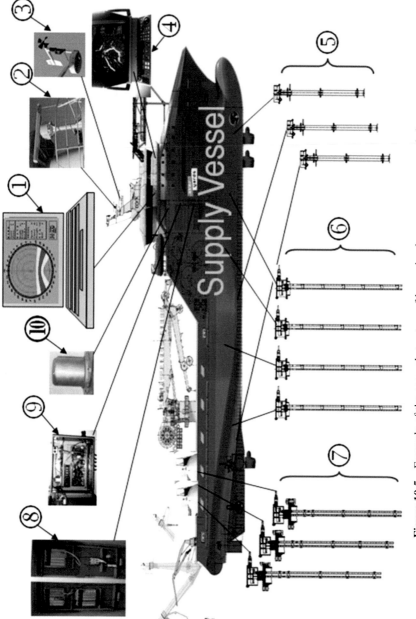

Figure 10.5 Example of the general structure of lascos hardware components and elements.

cargo quantity and quality monitoring and control (6); a set of polymetric sensors for liquefied LPG or LNG cargo quantity and quality monitoring and control (7); switchboards of the subsystem for actuating devices and operating mechanisms control (8); a basic electronic block of the subsystem for liquid, liquefied and loose cargo monitoring and control (9); a block with the sensors for real-time monitoring of ship dynamic parameters (10).

The structure of the software part of the typical sensory intelligent LASCOS is presented in Figure 10.6.

It consists of three main elements: a sensory monitoring agency (SMA) which includes three other sensory monitoring agencies – SSM (sea state, e.g. wind and wave model parameters), SPM (ship parameters) and NEM (navigation environment parameters); an information environment agency (INE) including fuzzy holonic models of ship state (VSM) and weather conditions (WCM), and also data (DB) and knowledge (KB) bases; and last but not least – an operator interface agency (OPIA) which provides the decision-making person (DMP) with necessary visual and digital information.

Unfortunately, until now, the above-mentioned challenges have not been combined together in an integrated model, which applies cutting-edge and novel simulating techniques. Agent-based computations are adaptive to information changes and disruptions, exhibit intelligence and are inherently distributed [4]. Holonic agents inherently may help design and operational

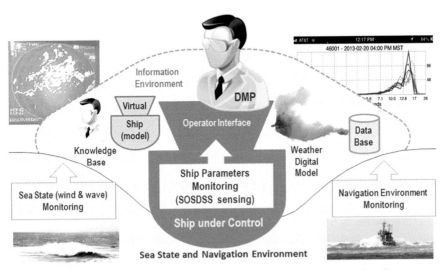

Figure 10.6 The general structure of lascos software elements and functions.

control processes in self-recovery and react to environmental real-time perturbations.

The agents are vital in a ship operations monitoring and control context, as ship safety refers to the inherently distributed and stochastic perturbation of its own state parameters and external weather excitations. Agents are welcome in the on-board LASCOS system design because they provide properties such as autonomy, responsiveness, distributiveness, openness and redundancy [15]. They can be designed to deal with uncertain and/or incomplete information and knowledge and this is extremely topical for fuzzy LASCOS as a whole. On the other hand, the problem of their sensing has never been under systematic consideration yet.

10.4.1.1 Sensory monitoring agency SMA

As mentioned above, all initial real-time information for the LASCOS as well as the sensory system is provided to DMP by the holonic agent-based sensory monitoring agency (SMA) which includes three other sensory monitoring agencies – SSM, SPM and NEM (see Figure 10.7).

The simultaneous functioning of all the structures presented in Figure 10.5, Figure 10.6 and Figure 10.7, i.e. hardware, logic and process algorithms, the corresponding software components and elements of LASCOS provide online forming of information environment for the system operator or seafarer – the decision-maker.

The polymetric sensing of the computer-aided control and monitoring system for cargo and ballast tanks is designed to ensure effective and safe control of ship operations in remote mode. This system is one of the most critical systems on-board a ship or any other ocean-floating vehicle. The operation of the system must be finely adjusted with the requirements of the marine industry. The polymetric system is a high-tech solution for the task of on-line monitoring and control of the ship liquid cargoes and ballast water tanks state. These systems are also designated for ship safety parameter monitoring and for docking operations control.

The solution provides safe and reliable operations in real-life harsh conditions, reduces risks for vessels and generates both operational and financial benefits for customers providing:

- The manual, automated or automatic remote control of cargo handling operations;

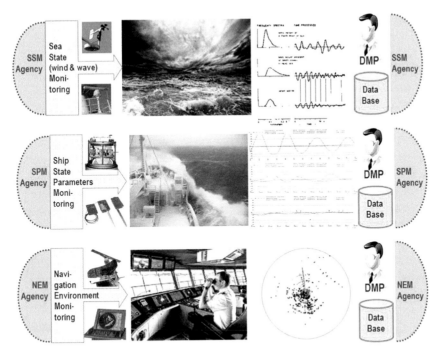

Figure 10.7   The General Structure of LASCOS Holonic agencies functions.

- The on-line monitoring of level, volume, temperature, mass, centre of mass for liquid, loose cargo or ballast water in all tanks;
- The monitoring and control of list and trim, draft line, bending and sagging lines;
- The monitoring of the vessel position during operation or docking; level indication and alarm of all monitored tanks;
- The remote control of cargo valves and pumps according to the actual operational conditions; the feeding of control signals to actuators, valves and pumps; the monitoring of hermetic dryness of dry tanks and conditions of water ingress in the possible damage conditions;
- The audible and visual warning about risky process parameters deviation from the stipulated values;
- The registration and storage of retrospective information on operations process parameters, equipment state and operator actions ("Black Box" functions);

- Comprehensive data and knowledge bases management;
- The testing and the diagnostics of all equipment operability.

10.4.1.2 Information Environment Agency INE

The information environment agency (INE) includes the fuzzy holonic models of ship state (VSM), the weather conditions model (WCM), and also data (DB) and knowledge (KB) bases. The central processor of the LASCOS system processes all the necessary initial information concerning ship design and structure in all possible loading conditions using the digital virtual ship model (VSM).

The information concerning the actual distribution of liquid and loose cargoes (including their masses and centres of gravity) are provided by the sensory monitoring agency (SMA). This information is combined with the information concerning actual weather and navigation environment details that are converted into the actual weather condition model (WCM). The information, permanently refreshable by SMA, is stored in the database and it is used in real-time mode by the VSM and WCM models [18].

10.4.1.3 Operator Interface Agency OPI

The operator interface agency (OPI) of the LASCOS system provides the decision-maker with all the necessary visual and digital information in the most convenient way for decision-making support on quality, efficiency and timeliness. Examples of some of these interfaces are presented in Figure 10.8 and Figure 10.9.

One of the functions of the LASCOS system is to predict safe combinations of ship speed and course in rough sea actual weather conditions. On the safe-storming diagram, the seafarer may see three coloured zones of "speed-course" combinations: the green zone of completely safe ship speed and course combination (grey zone in Figure 10.8); the dangerous red zone with "speed-course" combinations leading to ship capsizal with up to more than 95% probability (the black zone in Figure 10.8); and the yellow zone with "speed-course" combinations with intensive ship motions, preventing normal navigation and other operational activities (the white zone in Figure 10.8).

The centralized monitoring and control of ship's operational processes is performed via the operator interface from the main LASCOS system processor screen located in the wheelhouse. The advanced operator interface is playing an increasingly important role in the light of the visual management concept. The proposed operator interface for LASCOS provides all traditional man-machine control functionality. Moreover, this interface is clearly structured,

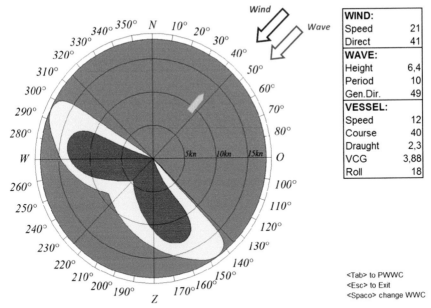

Figure 10.8 Safe Storming Diagram Interface of LASCOS.

powerful, ergonomic and easy to understand; it also makes the immediate forecast of ship behaviour under the control action chosen by the operator, thus preventing too risky and unsafe decisions. It also ensures all control processes rules and requirements consolidating all the controllable factors for better efficiency of operations to exclude human factor influence on decision-making. The interface also ensures system security via personal password profiles for nominated responsible and thoroughly trained persons to prevent the incorrect usage of equipment and to avoid the unsafe conditions of ship operations.

10.4.1.4 Advantages of the polymetric sensing

The described example of a sensory system for typical offshore supply vessel (presented in Figure 10.5) can be used for the demonstration of the polymetric sensing advantages.

The structure of the typical cargo sensory system for the offshore supply vessel consists of the following parts: a set of the sensors for the fuel-oil, liquefied LPG or LNG, ballast/drinking water and other liquid cargo quantity and quality monitoring; a set of sensors for the bulk cargo quantity and quality monitoring.

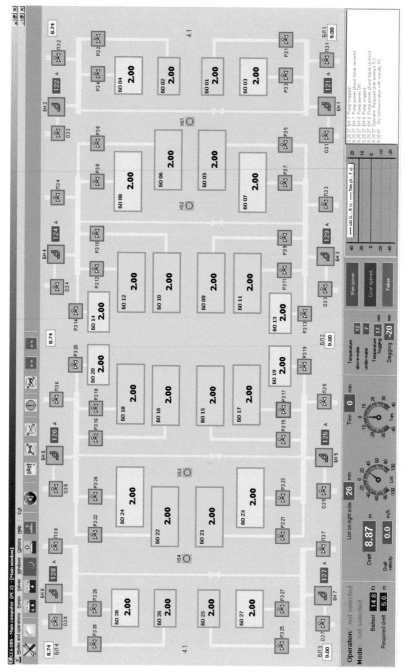

Figure 10.9   The Main Window of Ballast System Operations Interface.

Each tank is designated for the particular cargo type and equipped with the required sensors. For example, for the measurement of the diesel oil parameters, the corresponding tanks are equipped with the level sensors (level sensors with the separation level measurement feature) and temperature sensors. The total number of the tanks, corresponding sensors for the tanks of a typical supply vessel, are shown in Table 10.1 (in case of the traditional sensory system).

All the information acquired from the sensors must be pre-processed for the final calculation of the required cargo and ship parameters in the computing system. Each of the sensors requires power and communication lines, acquisition devices and/or interface transformers (e.g. current loop into RS-485 MODBUS) and so on.

In contrast to the classical systems, the polymetric sensory system requires only one sensor for the measurement of all required cargo parameters in the tank. Therefore, if we assume that traditional and polymetric sensory systems are equivalent in the measurement information quality and reliability (systems are interchangeable without any loss of measurement information quality), it is obvious that polymetric system has the advantage in the measurement channels number.

The cost criterion can be used for the comparison of the efficiency of traditional and polymetric sensory systems. Denoting the cost of one

Table 10.1 Quantity and Sensor Types for the Traditional Cargo Sensory System (Example)

Cargo Type	Measureable Parameters	Tanks Number	Sensors Number/Tank	Total Sensors Number
Diesel oil	Level in the tank, presence of water in the tank, temperature	6	3	18
LPG	Level in the tank, quantity of the liquid and vapor gas, temperature	1	3	3
Ballast water	Level in the tank	6	1	6
Drinking water	Level in the tank, presence of other liquids in the tank	6	2	12
Bulk cargo	Level in the tank, quality parameter (e.g. moisture content)	2	2	4
Total		21		43

measurement channel (the sensor + communication/power supply lines + transformers/transmitters) as C_{TMS} and C_{PMS}, the cost of the processing complex as C_{TPC} and C_{PPC}, the costs of the sensory systems for traditional sensory system C_{TSS} and for polymetric sensory system C_{PSS} can be estimated:

$$C_{TSS} = C_{TPC} + N_{TSS} \cdot C_{TMS},$$
$$C_{PSS} = C_{PPC} + N_{PSS} \cdot C_{PMS},$$

where N_{TSS}, N_{PSS} – the numbers of sensors used in the traditional and polymetric sensory system correspondingly.

Assuming that

$$C_{TPC} \approx C_{PPC} \approx C_{TMS} \approx C_{PMS},$$

the efficiency comparison of the polymetric and traditional sensory system $E_{TSS/PSS}$ can be roughly estimated using the costs of the equivalent sensory systems [8]:

$$E_{TSS/PSS} = C_{TSS}/C_{PSS} = \frac{N_{TSS} + 1}{N_{PSS} + 1}.$$

As can be seen from the above-described supply vessel example, the polymetric sensory system efficiency is two times greater than the efficiency of the traditional sensory system ($E_{TSS/PSS}$ = 44/22 = 2). It is worth mentioning that this comparison of the system efficiency is very rough and it is used only for the demonstration of the main advantage of the polymetric sensory system.

10.4.1.5 Floating dock operation control system

Another example of naval polymetric sensory systems is a computer-aided floating dock ballasting process control and monitoring system (CCS DBS) with the main interface window presented in Figure 10.9.

These systems were designated to ensure effective and safe control of docking operations in the automatic remote mode. The ballast system is one of the most critical systems on the floating dock. The operation must be finely coordinated with the requirements of the marine industry. The polymetric system enables to take high-tech solutions to support safe control and monitoring of the dock ballast system. This solution provides the safe and reliable operations of dock facilities in real-life harsh conditions, reduces risks for vessels and generates both operational and financial benefits to customers.

The main system interface window (see Figure 10.9) contains the technological equipment layout and displays the actual status of a rudder, valves, pumps, etc. The main window consists of the main menu, the toolbar, the information panel and also the technological layout containing control elements.

It is possible to start, change parameters or stop technological processes by clicking a particular control element. The user interface enables the operator to efficiently supervise and control every detail of any technological process. All information concerning ship safety and operation monitoring and control, event and alarm management, database management or message control is structured in functional windows.

10.4.1.6 Onshore applications

Another example of the successful polymetric sensing of computer-aided control and monitoring systems is connected with different marine terminals (crude oil, light and low-volatility fuel, diesel, liquefied petroleum gas, grain) and even with bilge water cleaning shops. Such sensing allows simultaneous monitoring of practically all the necessary parameters using one sensor for each tank, silo or other reservoir.

Namely, a set of the following characteristics is taken by a single polymetric sensor:

- The level, separation level of the non-mixed media and the online control of upper/lower critical levels and volume of the corresponding cargo in each tank;
- The temperature field in the media, the temperature of the product at particular points;
- Density, octane/cetane number of a fuel, propane-butane proportion in petroleum gas, presence and percentage of water in any mixture or solution (including aggressive chemicals – acids, alkalis, etc.);

As a result, a considerable increase of the measuring performance factor $E_{TSS/PSS}$ was achieved in each application system (up to 4–6 times [8]) with the essential concurrent reduction of number of instruments and measuring channels of the monitoring system. All the customers (more than 50 objects in Ukraine, Russia, Uzbekistan, etc.) reported commercial benefits after serial SADCO^{TM} systems deployment.

10.4.1.7 Special applications

Special applications of polymetric sensing of control and monitoring systems were developed for the following: real-time remote monitoring of motor lubrication oil production process (quantitative and qualitative control of the components depending on time and real temperature deviation during the production process); real-time remote control of aggressive chemicals at a nuclear power station water purification shop (quantitative and qualitative control of the components depending on time and real temperature deviation during the production process); water parameters control in the primary coolant circuit at a nuclear power station in normal and post-emergency operation state (fluidized bed-level control, pressure and temperature monitoring – all in the conditions of increased radioactivity).

10.5 Conclusions

This paper has been intended as a general presentation of a rapidly developing area of the promising transition from traditional on-board monitoring systems to intelligent sensory decision support and control systems based on neoteric polymetric measuring, data mining and holonic agent techniques. The area is especially attractive to those researchers who are attempting to develop the most effective intelligent systems by squeezing the maximum information from the simplest and most reliable sensors by means of sophisticated and effective algorithms.

In order for monitoring and control systems to become intelligent, not only for exhibition demonstrations and show presentations, but for real industrial applications, one needs to implement the leading-edge solutions for each and every component of such a system.

Combining polymetric sensing, data mining and holonic agencies techniques into one incorporated approach seems to be rather prospective if more and more research will develop appropriate theory models and integrate them into the practice as well.

References

[1] K. Hossine, M. Milton and B. Nimmo, 'Metrology for 2020s', Middlesex, UK: National Physical Laboratory, 2012, p. 28.
[2] S. Russell and P. Norvig, 'Artificial Intelligence: A Modern Approach', Upper Saddle River, NJ: Prentice Hall, 2003.
[3] L. Monostori, J. Vancza and S. Kumara, 'Agent-Based Systems for manufacturing', Annals of the CIRP, no. 55, 2 2006.
[4] P. Leitao, 'Agent–based distributed manufacturing control: A state-of-the-art survey', Engineering Applications of Artificial Intelligence, no. 22, pp. 979–991, 2009.

[5] D. McFarlane and S. Bussman, 'Developments in Holonic Production Planning and Control', Int. Journal of Production Planning and Control, vol. 6, no. 11, pp. 522–536, 2000.

[6] F. Amigoni, F. Brandolini, G. D'Antona, R. Ottoboni and M. Somal vico, 'Artificial Intelligence in Science of Measurements: From Measurement Instruments to Perceptive Agencies', IEEE Transactions on Instrumentation and Measurements, vol. 3, no. 52, pp. 716–723, 6 2003.

[7] Y. Zhukov, 'Theory of Polymetric Measurements', in Proceedings of 1st International Conference: Innovations in Shipbuilding and Offshore Technology, Nikolayev, 2010.

[8] Y. Zhukov, B. Gordeev and A. Zivenko, 'Polymetric Systems: Theory and Practice', Nikolayev: Atoll, 2012, p. 369.

[9] Y. Zhukov, 'Concept of Hypersignal and it's Application in Naval Cybernetics', in Proceedings of 3-rd International Conference: Innovations in Shipbuilding and Offshore Technology, Nikolayev, 2012.

[10] B. Marchenko and L. Scherbak, 'Foundations of Theory of Measurements', Proceedings of Institute of Electrodynamics of National Academy of Sciences of Ukraine, Electroenergetika, pp. 221–230, 1999.

[11] B. Marchenko and L. Scherbak, 'Modern Concept for Development of Theory of Measurements', Reports of National Academy of Sciences of Ukraine, no. 10, pp. 85–88, 1999.

[12] Y. Zhukov, 'Fuzzy Algorithms of Information Processing in Systems of Ship Impulse Polymetrics', in Proceedings of 1st International Conference: Problems of Energy saving and Ecology in Shipbuilding, Nikolayev, 1996.

[13] Y. Zhukov, 'Solitonic Models of Polymetric Monitoring Systems', in Proceedings of International Conference "Shipbuilding problems: state, ideas, solutions", USMTU, Nikolayev, 1997.

[14] Y. Zhukov, B. Gordeev and A. Leontiev, 'Concept of cloned Polymetric Signal and it's application in Monitoring and Expert Systems', in Proceedings of 3-rd International Conference on Marine Industry, Varna, 2001.

[15] Y. Zhukov, B. Gordeev and A. Zivenko, 'Polymetric sensing of intelligent robots', in Proceedings of IEEE 7th Int. Conf. on IDAACS, Berlin, 2013.

[16] A. Zivenko, A. Nakonechniy and D. Motorkin, 'Level measurement principles & sensors', in Materialy IX mezinarodni vedecko-prackticka conference "Veda a technologie: krok do budoucnosti - 2013", Prague, 2013.

[17] A. Zivenko, 'Forming and pre-processing of the polymetric signal', Collection of Scientific Publications of NUS, vol. 11, no. 439, pp. 114–122, 2011.

[18] Y. Zhukov, 'Instrumental Ship Dynamic Stability Control', in Proceedings of 5th IMAEM, Athens, 1990.

11

Design and Implementation of Wireless Sensor Network Based on Multilevel Femtocells for Home Monitoring

D. Popescu, G. Stamatescu, A. Măciucă and M. Struţu

Department of Automatic Control and Industrial Informatics, University "Politehnica" of Bucharest, Romania
Corresponding author: G. Stamatescu <grigore.stamatescu@upb.ro>

Abstract

An intelligent femtocell-based sensor network is proposed for home monitoring of elderly or people with chronic diseases. The femtocell is defined as a small sensor network which is placed into the patient's house and consists of both mobile and fixed sensors disposed on three layers. The first layer contains body sensors attached to the patient that monitor different health parameters, patient location, position and possible falls. The second layer is dedicated for ambient sensors and routing inside the cell. The third layer contains emergency ambient sensors that cover burglary events or toxic gas concentration, distributed by necessities. Cell implementation is based on the IRIS family of motes running the embedded software for resource-constrained devices, TinyOS. In order to reduce energy consumption and radiation level, adaptive rates of acquisition and communication are used. Experimental results within the system architecture are presented for a detailed analysis and validation.

Keywords: wireless sensor network, ambient monitoring, body sensor network, ambient assisted living.

11.1 Introduction

Recent developments in computing and communication systems applied to healthcare technology give us the possibility to implement a wide range of home-monitoring solutions for elderly or people with chronic diseases

Advances in Intelligent Robotics and Collaborative Automation, 235–256.

[1]. Thus, people may perform their daily activities while being constantly under the supervision of the medical personnel. The indoor environment is optimized such that the possibility of injury is minimal. Alarm triggers and smart algorithms sent data to the right intervention units in regard to the detected emergency [2].

When living inside closed spaces, small variables may be significant to the entire person's well-being. Therefore, the quality of the air, temperature, humidity or the amount of light inside the house may be important parameters [3]. Reduced costs, size and weight and energy-efficient operation of the monitoring nodes, together with the more versatile wireless communications, make the daily usage of the systems monitoring health parameters more convenient. By wearing them, patients are free to move at their own will inside the monitored perimeter, practically forgetting their presence. The goal is to design the entire system operation for a long period of time without human intervention and at the same time, triggering as few false alarms as possible.

Many studies investigated the feasibility of using several sensors placed on different parts of the body for continuous monitoring [4]. Home care for the elderly and chronic disease persons becomes an economic and social necessity. With a growing population of ageing people and the health care prices rising all over the world, we expect a great demand for home care systems [5, 6]. An Internet-based topology is proposed in [7] for the remote home-monitoring applications that use a broker server, managed by a service provider. The security risks from the home PC are transferred to the broker server and removed, as the broker server is located between the remote-monitoring devices and the patient's house. An early prototype of a mobile health service platform that was based on Body Area Networks is MobiHealth [8]. The most important requirements of the developer for an e-health application are size and power consumption, as considered in [9]. Also, in [10], a thorough comprehensive study of the energy conservation challenges in wireless sensor networks is carried out.

In [11], a wireless body area network providing long-term health monitoring of patients under natural physiological states without constraining their normal activities is presented.

Integrating the body sensors with the existing ambient monitoring network in order to provide a complete view of the monitored parameters is one of the issues discussed in this paper. Providing a localization system and a basic algorithm for event identification is also part of our strategy to fulfill all possible user requests. Caregivers also value information about the quality of air inside the living area. Many false health problems are usually related to the lack of oxygen or high levels of CO or CO_2.

The chapter is organized as follows: Section 2 provides an overall view on the proposed system architecture and detailed insight into the operation requirements for each of the three layers for body, ambient and emergency monitoring. Section 3 introduces the main criteria for efficient data collection and a proposal for an adaptive data rate algorithm for both the body sensor network and the ambient sensor network. This has the aim of reducing the amount of data generated within the networks, considering processing, storage, and energy requirements. Implementation details and experimental results are evaluated in Section 4, where the path is set for long-term deployment and validation of the system. Section 5 concludes the chapter and highlights the main directions for future work.

11.2 Network Architecture and Femtocell Structure

The proposed sensor network architecture is based on hybrid femtocells. A hybrid femtocell contains sensors which are grouped based on their functional requirements, mobility and energy consumption characteristics in three layers: the body sensor network (BSN), the ambient sensor network (ASN) and the emergency sensor network (ESN), as presented in Figure 11.1. Coordination is implemented through a central entity called the femtocell management node (FMN) which aggregates data from the three layers and acts as a interface to the outside world by means of the internet. Communication between different components can be made using wireless technology and radio compatible fiber optic. Table 11.1 lists the high-level characteristics of the two low-power wireless communication standards often used in home-monitoring scenarios: IEEE 802.15.1 and IEEE 802.15.4. This highlights an important design trade-off in the deployment of a wireless sensor network for home monitoring.

Table 11.1 Main characteristics of IEEE 802.15.1 and 802.15.4

IEEE Standard	802.15.1	802.15.4
Frequency	ISM - 2.4GHz	ISM/868/915 MHz
Data rate	1 Mbps	250 kbps
Topology	Star	Mesh
Scalability	Medium	High
Latency	Medium	High
Interference mit.	FHSS/DHSS	CSMA/CA
Trademark	Bluetooth	ZigBee

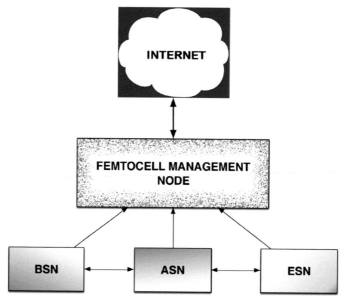

Figure 11.1　Hybrid femtocell configuration.

While IEEE 802.15.4 and ZigBee enable large dense networks with complex mesh topologies, the use of Bluetooth can become an advantage in applications with higher data-rate requirements and low latency.

The layer characteristics and functionalities are further elaborated upon.

11.2.1 Body Sensor Network

The body sensor network functional design includes battery charging nodes along with user-friendly construction and operation. In an idealised case, the size and weight would go unnoticed, immediately or after a short accomoda-tion period, not disturbing the patient or elderly person when wearing them. Nodes communicate using a low-power wireless communication protocol for very short distance data transmission and reception e.g. ZigBee or Bluetooth, depending on the data streaming rate of the application and energy resources on the node. Very low-energy consumption is an essential design criteria as changing the batteries on a daily basis becomes stressful on the long term and ideally the nodes would be embedded into wearable technology. Some on the sensed parameters for the BSN include dual- or tri- axial accelerom-eters, blood pressure, ECG, blood oxygen saturation, heart rate and body temperature.

The main events that should be detected by the BSN cover fall detection, activity recognition and variations in investigated parameters corresponding to alert and alarm levels.

11.2.2 Ambient Sensor Network

The ambient sensor network is comprised of a series of fixed measurement nodes, placed optimally in the target area/environment as to maximize sensing coverage and network connectivity through a minimum number of nodes. The low-power communication operates on longer communication links than the BSN and has to be robust to main phenomena affecting indoor wireless communication, such as interference, reflections and assymetric links. Though fixed node placement provides more flexibility when choosing the energy source, battery operation and low maintenance, enabled by low-energy consumption is preferred to mains power. For example, in the ASN, the router nodes which are tasked with redirecting much of the network traffic in the multi-hop architecture can be operated from the main power line.

Within the general framework, the ASN can serve as an external system for patient monitoring through a localization function based on link quality and signal strength. The monitored parameters include ambient temperature, humidity, barometric pressure and light. These can be evaluated individually or can serve as input data to a more advanced decision support system which can correlate the evolution of indoor parameters with the BSN data from the patient to infer the conditions for certain diseases. Some nodes of the ASN might include complex sensors like audio and video capture giving a more detailed insight into the patient's behaviour. Their current use is somewhat limited by the additional computing and communication ressources needed to accomodate the sensors into current wireless sensor network architectures as well as by privacy and data-protection concerns.

11.2.3 Emergency Sensor Network

The multi-level femtocell reserves a special role for the emergency sensor network which can be considered as a particular case of ASN tasked with quick detection and reaction to life or property threats through embedded detection mechanisms. The sensing nodes are fixed and their placement is well suited to the specifics of the measurement process. As an example, the house could be fitted with gas sensors in the kitchen next to a stove, carbon dioxide sensors would be placed in the bedroom and passive infrared sensor and pressure sensors would be fitted next to doors and windows. As the operation of the

ESN is considered critical, energy-efficient operation becomes a secondary design criteria. The nodes should have redundant energy sources, both batteries and mains power supply, and redundant communication links. For example, a wireless low latency communication protocol with simple network topology to minimize overhead and packet loss can be used as a main interface with the possibility of switching to a wired interface or using the ASN infrastructure in certain situations.

The main tasks of the ESN are to detect dangerous gas concentrations posing threats of explosion and/or intoxication and to detect intruders into the home such as burglars.

11.2.4 Higher-level Architecture and Functional Overview

One of the features of the hybrid femtocell is that the body sensor always interacts with the closest ambient sensor node, Figure 11.2, in order to send data to the gateway. This function reassures us that the sensors attached to the monitored person are always connected and the energy consumption is optimal because the distance between the person and the closest router is minimal. This feature can also be used as a mean of localization. The person wearing the

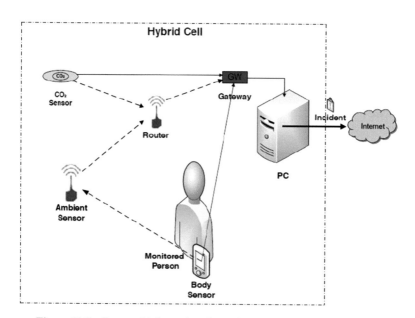

Figure 11.2   Data and information flow within and outside the femtocell.

body sensor will be always connected to the closest router. By using fixed environmental sensors with own ID and previously known positions, we can determine which room is presently used by the inhabitant. In order to have an accurate localization of the patient, an attenuation map of the sensors from each room must be created. Considering that patient is localization in a certain room of the home, by the closest ASN, is not accurate this could happen due to the following scenario: lets suppose that we have an ASN located in the bedroom, situated in the left side of the room. In the right part of the room, we have the door to the living room, and near this door we have the ASN for the living room. If the patient is located in the bedroom, but very close to the door, it will have as closest ASN the one from the living room, but he/she is situated in the bedroom. In order to avoid this localization error, we introduce the attenuation map of the sensors. Every ASN that localizes the BSN on the patient will transmit an attenuation factor. This way, using the attenuation factors from each sensor, we can localize the patient very accurately. In our example, if the bedroom ASN has a 10% factor, and the living room ASN has a 90% factor, using the attenuation map, we localize the patient as being in the bedroom, but very close to the door between the two rooms.

The position of a hybrid femtocell in the large wireless sensor network system is presented in Figure 11.3. Its main purpose is to monitor and interpret data, sending specific alarms when required. The communication between the femtocells and the network is based on internet. The same method is used for

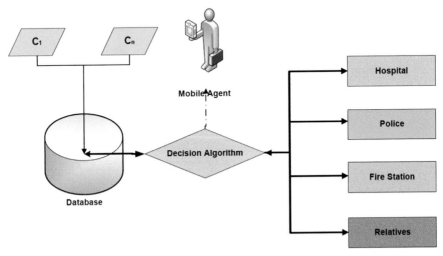

Figure 11.3 System architecture.

the communication between the network administrator and hospital, police, fire station or relatives.

In Figure 11.4, the ambient sensors' spatial placement and reference deployment inside the patients apartment is showcased. This fixed topology is consistent with the routing role of the ambient sensors in regard to the considered mobile body sensors. The goal of the system is to collect relevant data for reporting and processing. Therefore, achieving a very high sensory and communication coverage is among our main objectives.

The entire monitoring system benefits from a series of gas sensor modules strategically placed throughout the monitored space. Thus, the Parallax-embedded gas-sensing module for CO_2 is designed to allow a microcontroller to determine when a preset carbon dioxide gas level has been reached or exceeded. Interfacing with the sensor module is done through a 4-pin SIP header and requires two I/O pins from the host microcontroller. The sensor module is mainly intended to provide a means of comparing carbon dioxide sources and being able to set an alarm limit when the source becomes excessive [4].

A more advanced system for indoor gas monitoring, MICARES, is shown in Figure 11.5. It consists of an expansion module for a typical wireless sensor network platform with the ability to measure CO, CO_2 and O_3 concentrations and perform on board duty-cycling of the sensors for low energy as well as measurement compensation and calibration. Using this kind of platform, data

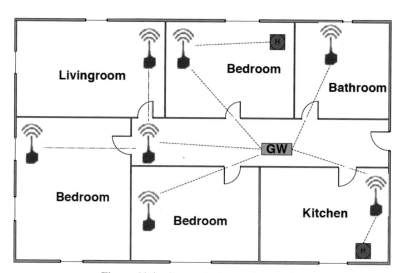

Figure 11.4   Sensor deployment example.

Figure 11.5 Embedded carbon dioxide sensor node module [12].

can be reliably evaluated locally and relayed to the gateway as small packets of timely information. The module can be either powered by mains or draw power from the energy source of the host node. Also, it can operate as a stand-alone device through ModBus serial communication with other information processing systems.

In case an alert is triggered, the information is automatically processed and the alarm is sent to the specific intervention factor (hospital, police, fire department, etc.). These have the option of remotely accesing the fem-tocell mangement node in order to verify that the alarm is valid and act by dispatching an intervention team to solve the issue. Subsequently, all the alarms generated over time are classified and stored in a central database for reporting purposes.

11.3 Data Processing

Data harvesting and processing is performed at the femtocell level, so the whole network is not flooded by useless data from all femtocells. Considering the main goals of conserving node energy, while taking into account the limited storage and processing capabilities of each node, we proposed a strategy to dynamically adapt the sampling rate of the investigated parameters for the BSN and ASN. The ESN is considered crtical home infrastructure and it is not affected by this strategy, with its own specific challenges for reliability, periodic sensor maintenance and calibration.

Adaptive data acquisition is implemented by framing collected data into a safety interval given by [V_{min}, V_{max}]. While the raw values are relatively close to the interval center $V = \frac{1}{2}(V_{max} - V_{min})$ and the recent variation of the time series, given by the derivative, is low over a exponential weighted time horizon, then the data acquisition rate is lowered periodically towards a lower bound. When collected measurements start to vary, data acquisition is increased steeply in order to capture the significant evene, then a stability period is observed and it begins to be lowered again. Though local data storage on the node can be exploited, we prefer to synchornize collection with transmission of the packets through the radio interface in order to conserve data freshness which can prove important in order to effectively react to current events. Energy level must be considered in order to prevent the reception of corrupted data and to avoid unbalanced node battery depletion e.g. in the case of routing nodes in the network. This represents a validation factor of the data and is automatically transmitted together with the parameter value. Energy-aware policies can be implemented at the femotcell level to optimize battery recharge, in the BSN case, and battery replacement, for the ASN.

Evaluation of the parameter derivate takes place for attaining a variable rate of acquisition. The values obtained by calculating the derivate also help us to decide what type of event has happened. This can be done by building a knowledge base during system commissioning and initial operating period which can be used to relate modifications in observed parameter to trained event classes. These can go from trivial associations like high temperature and low oxygen/smoke meaning a fire, to subtle patient state of health detection by changing vital signs and aversion to light. The algorithm described can be summarized as follows:

Data: measured values, measurement derivative, energy level
Result: adaptive data rate initialize;
while *network is operational* **do**
> collect measurement values;
> check distance from interval center;
> compute time series derivative;
> **if** *high sudden variation* **then**
> > increase data rate;
>
> **else**
> > lower data rate;
>
> **end**

end

Algorithm 1: Adaptive Data Collection for Home Monitoring

11.4 **Experimental Results**

In order to bring value to the theoretical system architecture proposed, two experiments have be devised and implemented. They cover the body and ambient sensor layers of the multi-level femtocell for home monitoring. The main operational scenario that was considered involves an elderly patient, living alone at home in an appartment or house with multiple rooms. A caregiver is available on call as well as a permanent connection to the emergency services exists, with the possibility on alerting close relatives in the process. As functional requirements, we target activity monitoring and classification by body-worn accelerometers and ambient measurement of humidity, temperature, pressure and light, along with their correlations.

The first experiment is performed as part of the body sensor network. Therefore, two accelerometers with two axes were placed on the patient, one on the right knee and the other one on his left hip, as shown in Figure 11.6. The sensors are automatically assigned a unique value ID. Therefore, in our

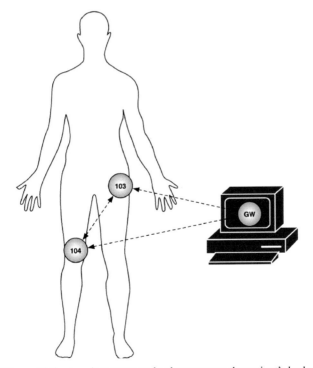

Figure 11.6 Accelerometer node placement on the patient's body.

experiment, the sensor situated on the knee has ID104, and the other one, placed on the left hip, ID103. In order to overcome the hardware limitations, in a tridimensional representation, the representation of the 3 axis using those two accelerometers is the following:

- X axis: front back, X axis of node 104;
- Y axis: left right, X axis of node 103;
- Z axis: bottom up, Y axis of both nodes.

The following activities were executed during the experiment: standing, sitting on a chair, standing again and slow walking from bedroom to living. Data acquisition has been performed using MOTE-VIEW [13]. This is an interface, client layer, between a user and a deployed network of wireless sensors. Besides this main function, the user can change or update the individual node firmware, switch from low-power to high-power mode and set the radio transmit power. Collected data is stored in a local database and can be accessed remotely by authorized third parties.

Multiple experiments have been conducted in order to determine the trust values. Therefore, in the X and Y axis charts presented in Figures 11.7 and 11.8, the readings outside the green lines represent that an event occurred, obtained by thresholding. The following events have been monitored during our experiment in this order: standing, sitting on a chair,

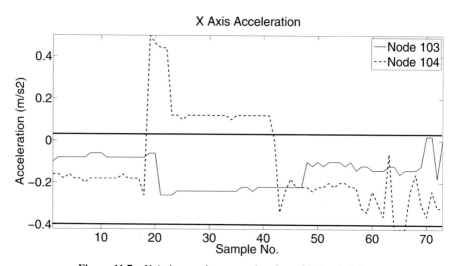

Figure 11.7   X Axis experiment acceleration with thresholding.

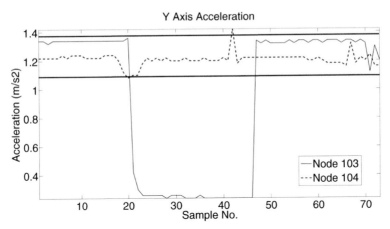

Figure 11.8 Y Axis experiment acceleration with thresholding.

standing again and slow walking from bedroom to living. The used sensors are ADXL202E [14].

These are low-cost, low-power, complete 2-axis accelerometer with a digital output, all on a single monolithic IC. They are an improved version of the ADXL202AQC/JQC. The ADXL202E will measure accelerations with a full-scale range of 2 g. The ADXL202E can measure both dynamic acceleration (e.g., vibration) and static acceleration (e.g., gravity). The outputs are analog voltage or digital signals whose duty cycles (ratio of pulse width to period) are proportional to acceleration. The duty cycle outputs can be directly measured by a micro- processor counter, without an A/D converter or glue logic. The duty cycle period is adjustable from 0.5 ms to 10 ms via a single resistor (RSET).

The radio base station is made up of an IRIS radio/processor board connected to a MIB520 USB interface board via the 51-pin expansion connector. The interface board uses a FTDI chip and provides two virtual COM ports to the host system. COMx is used for programming the connected mote and COMx+1 is used by middleware applications to read serial data. The used network protocol is called XMesh, owned by Crossbow, based on the standard 802.15.4. The Crossbow sensor networks can run different power strategies, each of these strategies being a trade-off between power, data rates and latency.

XMesh-HP [15] is the best approach for systems that have continuous power, offering the highest message rate, usually proportional to the band rate of the radio. Radios and processors are continually powered, consuming

Table 11.2 XMesh power configuration matrix

Power Mode	MICA2	MICA2DOT	MICAz
XMesh-HP	X	X	X
XMesh-LP	X	X	Async. only
XMesh-ELP	X	X	X

Table 11.3 XMesh performance summary table

Parameter	XMesh-HP	XMesh-LP	XMesh-ELP
Route update interval	36 sec.	360 sec.	36 sec. (HP) / 360 sec. (LP)
Data message rate	10 sec. typ.	180 sec., typ.	N/A
Mesh formation time	2–3 RUI	X	X
Average current usage	20–30 mA	400 uA	50 uA

between 15 and 30 mA depending on the type of the Mote. Route Update Messages and Health Messages are sent at a faster rate which decreases the time it takes to form a mesh or for a new mote to join the mesh.

XMeshLP [15] is used for battery-operated systems that require multi-month or multi-year life. It can run either time-synchronized or asynchronous. The best power efficiency is achieved with time synchronization within 1 msec. The motes wake-up, typically 8 times/second, time synchronized, for a very short interval to see if the radio is detecting any signal over the noise background. If this happens, the radio is kept alive to receive the signal. This action usually results in a base level current of 80 μA. The total average current depends on the number of messages received and transmitted. If data is transmitted every three minutes, the usual power in a 50-node mesh is around 220 μA. XMesh-LP can be configured for even lower power by reducing the wake-up periods and transmitting at lower rates. Also, route update intervals are set at a lower rate to conserve power, resulting in a longer mesh formation time.

XMesh-ELP [15] is only used for leaf nodes that communicate with parent nodes running XMesh-HP. A leaf node is defined as a node that does not participate in the mesh; it never routes messages from child motes to parent motes. The results of the ELP version are very low power because the mote does not need to use the time synchronized wake-up mode to check for radio messages. The mote can sleep for very long times, this way maintaining its neighborhood list to remember which parents it can select. If it does not get a link-level acknowledgement when it transmits to a parent, it will find another parent and so on. This operation can happen very quickly or might take some time if the RF environment or mesh configuration has changed considerably.

Because of their small size, nodes can be easily concealed into the background, interfering as little as possible with the user's day-to-day routine. We have also the possibility to set the sampling rate at a suitable level in order to achieve low-power consumption and by this a long operating range without human intervention [16].

Our infrastructure also offers routing facilities increasing the reliability of our network by self-configuring into a multihop communication system whenever direct links are not possible. After experimenting with different topologies, we achieved a working test scenario which consisted in a four-level multihop communication network which is more than we expect to be necessary in any of our deployment locations.

Extensive experimental work has been carried out for the ambient sensor layer of the system based on MTS400 IRIS sensor nodes. One of the reasons for choosing this specific board has been that it provided the needed sensors in order to gather a variety of environmental parameters, like temperature, humidity, relative pressure and ambient light. The experimental deployment consists of three measurement nodes organized in a true mesh topology in a testing indoor environment. These aim at modeling a real implementation in the patient's home and were taken over the course of a week-long deployment. In order to assure accounting to uneven sampling from the sensor nodes, we use as reference time MATLAB serial time units which are converted from conventional time stamps entries into the MOTE-VIEW database of the form *dd.mm.yyyy HH:MM:SS.*

In Figure 11.9(a), the evolution of the humidity parameter measured by indoor deployed sensors can be seen. The differences account for node placement in the different rooms and exposure to windows and doors. Subsequent processing can lead to computing average values and to other correlations with ambient and body parameters and an intelligent information system which can associate variations in ambient humidity and temperature to influences on chronic disease. Figure 11.9(b) illustrates temperature variations obtained from the ambient sensor network. These reflect the circadian evolution of the measured parameter and show the importance of correct node placement and data aggregation within the sensor network.

Barometric pressure (Figure 11.10(a)) is also observed by the sensor network over the testing period. This is the parameter that is least influenced by node placement and more by general weather trends. As differences between individual sensor node values of a few percentage points, these can be attributed to sensing element calibration or local temperature compensation. Aggregating data also in this case can lead to higher-quality measurements.

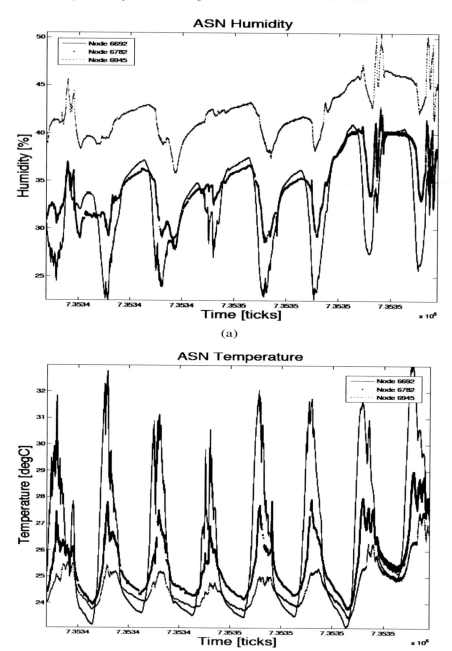

Figure 11.9   Measurement data: humidity (a); temperature (b).

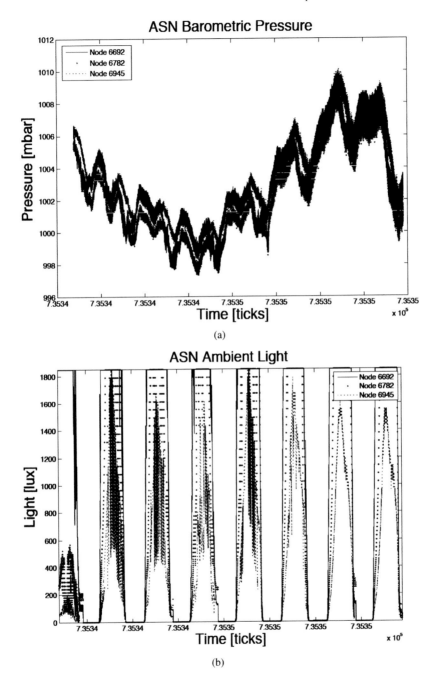

Figure 11.10   Measurement data: barometric pressure (a); light (b).

Ambiental light values, suitable for feeding data to an intelligent mood lighting system, are shown in Figure 11.10(b). The light sensor saturates in full daylight at around 1850 lux and quickly responds to variations in the measured light. The most important period of the day are dawn and dusk where the information provided can assure a smooth transition from artificial to natural light and reverse.

Data flow coming from the ambient sensor network is processed in multiple steps: at the source node, in the network e.g. through aggregation or sensor fusion, and at the home server or gateway level. Each steps converts raw data to higher-level pieces of information which can be more efficiently operated with and become meaninful through correlation and interactive visualization. Efficient management of this infomation is critical to correct operation of the home-monitoring system. Alerts and alarms have to be reliable and build trust among the end user leading to widespread acceptance whilst assuring a high level of integrity, security and user privacy.

Table 11.4 summarizes the aggregated experimental deployment values for the three nodes over the investigated period. Minimum, maximum, average and the standard deviations of the collected time series for each of the measured ambiental parameters are listed.

Making effective use of the large quantities of data generated by the layers of the femtocell structure, represented by the individual sensor networks, poses a significant challenge. One idea is to apply computational intelligence

Table 11.4   Ambiental Monitoring Summary

Node ID		6692	6782	6945
Humidity [%]	*min*	22.5	27.8	35.8
	max	41	42.8	50.5
	avg	33.39	34.62	42.26
	stdev	4.76	3.22	2.58
Temperature [degC]	*min*	23.08	23.78	23.4
	max	32.98	29.01	27.48
	avg	26.24	25.45	24.66
	stdev	2.7	1.1	0.75
Pressure [mbar]	*min*	997.83	997.34	998.24
	max	1009.6	1009.3	1010.2
	avg	1003.3	1002.2	1003.5
	stdev	2.85	2.7	2.7
Light [lux]	*min*	0	0	0
	max	1847.1	1847.1	1847.1
	avg	952.18	705.87	414.05
	stdev	880.32	801.35	517.96

techniques, either in a centralized manner at the gateway level or in a distributed fashion where computing taks are spread among the nodes according to a dynamic strategy. An example is given for time series prediction on the temperature data collected by the ambient sensor network using neural networks. As a tool, the MATLAB technical computing environment can be used for modeling, testing and validation and embedded code generation at the gateway level.

11.5 Conclusion

The chapter introduced a system architecture composed of a smart hybrid sensor network for indoor monitoring using a multilayer femtocell for ubiquitous intelligent home monitoring. The main components of the system are three low-level wireless sensor networks: body, ambient and emergency, along with a central coordination entity named the femtocell management node. This also acts as gateway towards the internet and the interested stakeholders in an ambient-assisted living scenario. It has been argued that efficient data collection and processing strategies along with robust networking protocols can enable seemless integration into the patient's home and bring added value to home care whilst reducing overall medical and assitance costs. Recent advances in miniaturization of discrete electronic components and systems, along with enhanced computing and communication capabilities of intelligent home device offer a good opportunity in this application area. This can be exploited for both reasearch and development in the field of ambient-assisted living to increase quality of life while dealing with increased medical costs.

The main experimental focus was on the body sensor network layer and the ambient sensor layer and experimental deployment and implementation has been illustrated. First, body-worn wireless accelerometers were used to detect and classify human activity based on time-domain thresholding. Second, extensive results from a medium-term deployment of an ambient sensor network were illustrated and discussed. This had, as a main purpose, the collection of ambient parameters, like temperature, humidity, barometric pressure and ambient light while observing network protocol behaviour. The main conclusion is that wireless sensor network systems and protocols offer a reliable option for deployment in home monitoring, given the specific challenges of indoor environments.

Plans for future development have been established on three main paths. One direction includes extending the system with more measured parameters through additional sensor integration with the wireless nodes. The focus here would be on the body sensor network side where a deep insight into the patient's well-being and health status can be gained. Also, while raw data and machine-learning algorithms can provide high-confidence recommendations and alerts to the caregivers, data visualization in the home and for the patient should not be neglected. This can be done by developing adaptive and intuitive interfaces for the patient or elderly person which enhance acceptance of the system. The quality and accuracy of the expected results has to be increased by integrating state-of-the-art sensing, signal processing and embedded computing hardware along with the implementation of advanced methods for experimental data processing.

References

[1] A. Măciucă, D. Popescu, M. Struu, and G. Stamatescu. Wireless sensor network based on multilevel femtocells for home monitoring. In 7th IEEE International Conference on Intelligent Data Acquisition and Advanced Computing Systems: Technology and Applications, pages 499–503, 2013.

[2] M. Fahim, I. Fatima, Sungyoung Lee, and Young-Koo Lee. Daily life activity tracking application for smart homes using android smart-phone. In Advanced Communication Technology (ICACT), 2012 14th International Conference on, pages 241–245, 2012.

[3] M. Smolen, K. Czopek, and P. Augustyniak. Non-invasive sensors based human state in nightlong sleep analysis for home-care. In Computing in Cardiology, 2010, pages 45–48, 2010.

[4] Yibin Hou, Na Li, and Zhangqin Huang. Triaxial accelerometer-based real time fall event detection. In Information Society (i-Society), 2012 International Conference on, pages 386–390, 2012.

[5] Andrew D. Jurik and Alfred C. Weaver. Remote medical monitoring. Computer, 41 (4):96–99, 2008.

[6] P. Campbell. Population projections: States, 1995–2025. Technical report, U.S. Census Bureau, 1997.

[7] Chao-Hung Lin, Shuenn-Tsong Young, and Te-Son Kuo. A remote data access architecture for home-monitoring health-care applications. Medical Engineering and Physics, 29(2):199–204, 2007.

[8] Aart Van Halteren, Dimitri Konstantas, Richard Bults, Katarzyna Wac, Nicolai Dokovsky, George Koprinkov, Val Jones, and Ing Widya. Mobi-health: ambulant patient monitoring over next generation public wireless

networks. In in Demiris, G. (Ed.): E-health: Current Status and Future Trends, IOS, pages 107–122. Press, 2004.

[9] DD Vergados, D Vouyioukas, NA Pantazis, I Anagnostopoulos, DJ Vergados, and V Loumos. Decision support algorithms and optimization techniques for personal homecare environment. IEEE IntÄôl. Special Topic Conf. Information Technology in Biomedicine (ITAB 2006), 2006.

[10] S. Tarannum. Energy conservation challenges in wireless sensor networks: A comprehensive study. Wireless Sensor Network, 2010.

[11] Emil Jovanov, Aleksandar Milenkovic, Chris Otto, and PietC de Groen. A wireless body area network of intelligent motion sensors for computer assisted physical rehabilitation. Journal of NeuroEngineering and Rehabilitation, 2(1):1–10, 2005.

[12] G. Stamatescu, D. Popescu, and V. Sgarciu. Micares: Mote expansion module for gas sensing applications. In Intelligent Data Acquisition and Advanced Computing Systems (IDAACS), 2013 IEEE 7th International Conference on, volume 01, pages 504–508, Sept 2013.

[13] M. Turon. Mote-view: A sensor network monitoring and management tool. In Embedded Networked Sensors, 2005. EmNetS-II. The Second IEEE Workshop on, May 2005.

[14] T. Al-ani, Quynh Trang Le Ba, and E. Monacelli. On-line automatic detection of human activity in home using wavelet and hidden markov models scilab toolkits. In Control Applications, 2007. CCA 2007., Oct 2007.

[15] Crossbow Inc. XMesh User's Manual, revision c edition, March 2007.

[16] G. Stamatescu and V. Sgarciu. Evaluation of wireless sensor network monitoring for indoor spaces. In Instrumentation Measurement, Sensor Network and Automation (IMSNA), 2012 International Symposium on, Aug 2012.

[17] Parallax Inc. Co2 gas sensor module (27929) datasheet. Technical report, Parallax Inc., 2010.

12

Common Framework Model for Multi-Purpose Underwater Data Collection Devices Deployed with Remotely Operated Vehicles

M.C. Caraivan[1], V. Dache[2] and V. Sgarciu[2]

[1]Faculty of Applied Sciences and Engineering, University Ovidius
of Constanta, Romania
[2]Faculty of Automatic Control and Computers, University Politehnica
of Bucharest, Romania
Corresponding author: M.C. Caraivan <caraivanmitrut@gmail.com>

Abstract

This paper is following the development of real-time applications for marine operations focusing on modern modelling and simulation methods and presents a common framework model for multi-purpose underwater sensors used for offshore exploration. It is addressing deployment challenges of underwater sensor networks called by the authors "Safe-Nets" by using Remotely Operated Vehicles (ROV).

Keywords: Remotely Operated Vehicles, ROV, simulation, testing, object modelling, underwater component, oceanographic data collection, pollution.

12.1 Introduction

The natural disaster following the explosion of BP Deepwater Horizon offshore oil-drilling rig in the Gulf of Mexico has raised questions more than ever about the safety of mankind's offshore oil-quests. For three months in 2010, almost 5 million barrels of crude oil formed the largest accidental

Advances in Intelligent Robotics and Collaborative Automation, 257–294.

marine oil spill in the history of the petroleum industry. The frequency of maritime disasters and their effects appear to have dramatically increased during the last century [1], and this draws considerable attention from decision makers in communities and governments. Disaster management requires the collaboration of several management organizations resulting in heterogeneous systems. Interoperability of these systems is fundamental in order to assure effective collaboration between different organizations.

Research efforts in the exploration of offshore resources have increased more and more during the last decades, thus contributing to greater global interest in the area of underwater technologies. Underwater sensor networks are going to become in the nearby future the background infrastructure for applications which will enable geological prospection, pollution monitoring, and oceanographic data collection. Furthermore, these data collection networks could in fact improve offshore exploration control by replacing the on-site instrumentation data systems used today in the oil-industry nearby well heads or in well-control operations, e.g. using underwater webcams which can provide important visual data aid for surveys or for offshore drilling explorations. These facts lead to the idea of deploying multi-purpose underwater sensor networks along-side with oil companies' offshore operations. The study is trying to show the collateral benefits of deploying such underwater sensor networks and we address state-of-the-art ideas and possible implementations of different applications like military surveillance of coastal areas, assisting navigation [2] or disaster prevention systems – including earthquakes and tsunami detection warning alarms in advance – all in order to overcome the biggest challenge of development: the cost of implementation.

It is instructive to compare current terrestrial sensor network practices to underwater approaches: terrestrial networks emphasize low-cost nodes (around a maximum of US$100), dense deployments (at most a few 100m apart) and multi-hop short-range communication. By comparison, typical underwater wireless communications today are expensive (US$10.000 per node or even more), sparsely deployed (a few nodes, placed kilometres apart), typically communicating directly to a "base-station" over long-distance ranges rather than with each other. We seek to reverse the design points which make land networks so practical and easy to expand and develop, so underwater sensor nodes that can be inexpensive, densely deployed, and communicating peer-to-peer [3].

Multiple Unmanned or Autonomous Underwater Vehicles (UUVs, AUVs), equipped with underwater sensors, will also find application in exploration of natural undersea resources and gathering of scientific data in collaborative monitoring missions. To make these applications viable, there is a need to enable underwater communications among underwater devices. Ocean Sampling Networks have been experimented in the Monterey Bay area, where networks of sensors and AUVs, such as the Odyssey-class AUVs, performed synoptic, cooperative adaptive sampling of the 3D coastal ocean environment [4]. While offshore constructions' number grows, we should be able to implement auxiliary systems that allow us to better understand and protect the ocean surface we are building on. We will be using Remotely Operated Vehicles (ROVs) and a VMAX-Perry Slingsby ROV Simulator with scenario development capabilities to determine the most efficient way of deploying our underwater sensor networks, which we called "Safe-Nets", around offshore oil drilling operations, including all types of jackets, jack-ups, platforms, spars or any other offshore metallic or concrete structure.

The ability to have small devices physically distributed near offshore oil-fields' operations brings new opportunities to observe and monitor micro-habitats [5], structural monitoring [6] or wide-area environmental systems [7]. We even began to imagine a scenario where we can expand these sensor networks in order to slowly and steadily develop a global "WaterNet", which could be an extension of the Internet on land. In the same manner which allowed Internet networks on land to develop by constantly adding more and more nodes to the network, we could allow information to be transmitted from buoy to buoy in an access-point like system. These small underwater "Safe-Nets" could be joined together and the network could expand into a global "Water-Net" in the future, allowing data to be sent and received to and from shore bases. Of course, today, we can see considerable less kilobytes of data to be sent and received at first, but the main advantages would be in favour of disaster prevention systems. "Safe-Nets" for seismic activity and tsunami warning systems alone can represent one of the reasons for underwater network deployment, which are quite limited today compared to their counterparts on land. We propose a model of interoperability in case of a marine pollution disaster for a management system based upon Enterprise Architecture Principles.

If we keep in mind that the sea is a harsh environment, where reliability, redundancy and maintenance-free equipment are most desirable objectives, we should seek the methods and procedures for keeping the future development in a framework that should be backwards compatible with any other sensor nodes already deployed. In order to comply with the active need for upgrading to future technologies, we have thought of a common framework model with multiple layers and drawers for components which can be used for different purposes, but mainly for underwater data collection and monitoring. This development using Enterprise Architecture Principles is sustainable through time, as it is backed up by different solutions to our research challenges, such as power supply problem, fouling corrosion, self-configuration, self-troubleshooting protocols, communication protocols and hardware methods.

Two-thirds of the surface of Earth is covered by water and as history proved it, there is a constantly increasing number of ideas to use this space. One of the most recent is perhaps moving entire buildings of servers - Google's Data-Centres [8] overseas, literally, because of their cooling needs which nowadays are tremendous. These produce a heat footprint clearly visible even from satellites and by transferring them to the offshore environment, their overheating problems would have cheaper cooling methods which could be satisfied by the ocean's seawater almost constant temperature. Also, we discuss the electrical power supply possibilities further in the following chapter.

12.2 Research Challenges

We seek to overcome each of the design challenges that prohibited underwater sensor network development, especially by designing a common framework with different option modules available to be installed. If having a hardwire link at hand, by attaching these devices to offshore construction sites or to autonomous buoys, we could provide inexpensive sensors by using the power supply or communication means from that specific structure. We are looking forward to develop a variety of option modules for our common framework to be used for all types of underwater operations, which can include the instrumentation necessities nearby wellheads and drill strings or any type of oceanographic data collection, therefore becoming a solution at hand for any given task. This could provide the financial means of deploying underwater Safe-Nets, especially by tethering to all the offshore structures or exploration facilities which need different underwater data collection by their default nature.

12.2.1 Power Supply

Until now, only battery power was mainly used in underwater-based sensor deployments. The sensors were deployed and shortly afterwards were recovered. In our case, the close proximity to oil-rig platforms or other offshore constructions means already existing external power sources: diesel or gas generators, wind turbines, gas pressure turbines. We can overcome this design issue with cable connections to jackets or to autonomous buoys with solar panels which are currently undergoing research [9, 10].

Industrial applications such as oil-fields and production lines use extensive instrumentation, sometimes with the need of a video-feedback from the underwater operations site. Considering the depths at which these cameras should operate, there is also an imperative need for proper lighting of the area; therefore we can anticipate that these nodes will be tethered in order to have a power source at hand.

Battery power problems which in our case can be overcome not only by sleep-awake energy efficient protocols [11–13], but also by having connectivity at hand to other future system types of producing electricity from renewable resources, like wave energy converter units according to the European project Aquatic Renewable Energy Technologies (Aqua-RET) [14]:

- Attenuator-type Figure 12.1: Pelamis Wave Energy Converter [15];

- Axial symmetric absorption points as in Figure 12.2: WaveBob [16], AquaBuoy, OE Buoys [17] or Powerbuoy [18];

- Wave-level oscillation converters: completely submerged Waveroller or surface Oyster [19];
- Overtopping devices Figure 12.3: Wave Dragon [20];

Figure 12.1 Pelamis wave converter Orkney, U.K.

- Submersible differential pressure devices Figure 12.4: Archimedes Waveswing [21];

- Oscillating Water Column (OWC) devices.

Figure 12.2   Axial symmetric absorption buoy.

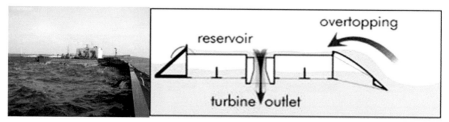

Figure 12.3   Wave Dragon - Overtopping devices principle.

Figure 12.4   Archimedes Waveswing (AWS).

Figure 12.5 Wind farms in North Sea.

In addition, we are considering also other types of clean energy technology production systems at sea:

- Wind farms: usually the wind speed at sea is far greater than on land, however, by comparison to its land counterpart, offshore wind turbines are harder to install and need more technical and financial efforts. The distance to land, water depth and sea floor structure are factors that need to be taken into consideration for Aeolian projects at sea. The first project for an offshore wind farm was developed in Denmark in 1991;
- Oceans' thermic energy by using the temperature difference between surface and depth waters, which needs to be at least 20°C at less than 100m from sea surface. These desiderates are usually full-filled nearby in Equatorial regions;
- Tidal waves and ocean currents such as Gulf Stream, Florida Straits, North Atlantic Drift possess energy which can be extracted with underwater turbines.

Besides the power supply facilities, all these devices themselves could in fact be areas of interest for deployment of our Safe-Net sensors.

12.2.2 Communications

Until now, there were several attempts to deploy underwater sensors that record data during their mission, but they were always recovered afterwards. This did not give the flexibility needed for real-time monitoring situations like surveillance or environmental and seismic monitoring. The recorded data could not be accessed until the instruments were recovered. It was also impossible to detect failures before the retrieval and this could easily lead to the complete failure of a mission. Also, the amount of data stored was limited by the capacity of the devices on-board the sensors (flash memories, hard disks).

Two possible implementations are buoys with high-speed RF-based communications, or wired connections to some sensor nodes. The communication bandwidth can be provided also by satellite connections which are usually present on offshore facilities. If linked to an autonomous buoy, the device provides GPS telemetry and has communication capabilities of its own. Therefore, once the information gets to the surface, radio communications are considered to be already provided as standard. Regarding underwater communications, usually the typical physical layer technology implies acoustic communications. Radio waves have long-distance propagation issues through sea water and can only be done at extra low frequencies, below 300 Hz [22]. This requires large antennae and high transmissions power, which we would prefer avoiding. Optical waves do not suffer from such high attenuation but are affected by scattering. Moreover, transmission of optical signals requires high precision in pointing the narrow laser beams. The primary advantage of this type of data transmission is the higher theoretical rate of transmission, while the disadvantages are the range and the line-of-sight operation needed. We did not consider this as a feasible solution due to marine snow, non-uniform illumination issues and other possible interferences.

We do not intend to mix different communication protocols with different physical layers, but we analyze the compatibility of each with existing underwater acoustic communications, state-of-the-art protocols and routing algorithms. Our approach will be a hybrid system, like the one in Figure 12.6 that will incorporate both tethered sensors and wireless acoustic where absolutely no other solution can be implemented (e.g.: a group of bottom sea floor anchored sensor nodes are implemented nearby an oil pipe, interconnected to one or more underwater "sinks", which are in charge of relaying data from the ocean bottom network the a surface station [23].

Regarding the propagation of acoustic waves in the frequency gamma we are interested in, for the multi-level communication between Safe-Net sensor nodes, we are looking into already known models [24]. One of the major problems related to the fluid dynamics are the non-linear movement equations, which imply the fact that there isn't a general exact solution. Acoustics represent the first order of approximation in which the non-linear

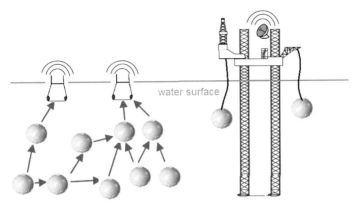

Figure 12.6 Possible underwater sensor network deployment nearby Jack-up rig.

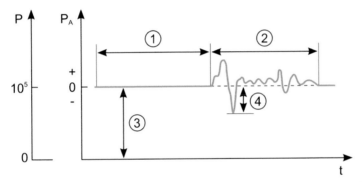

Figure 12.7 Sound pressure diagram: 1– Equilibrium; 2– Sound; 3– Environment Pressure; 4– Instantaneous Pressure of Sound.

effects are neglected [25]. Acoustic waves propagate because of the medium compressibility and the acoustic pressure or the sound pressure represents the local deviation of the pressure whose root cause can be traced back to a sound wave generated against the local environment. In air, the sound pressure can be measured using a microphone, while in water it can be measured using a hydrophone.

Considering the case of acoustic waves propagation in real fluids for our mathematic general formalism, we have made the following assumptions: gravity forces can be neglected, so equilibrium pressure and density get uniform values all over the fluid's volume (p_0 and ρ_0); the dissipative effects such as viscosity and thermic conductibility are negligible; the medium is homogenous, isotropic and has perfect elasticity, as well as the fluid particles speed is slow (the "*small amplitudes*" assumption). Therefore, we can write a Taylor development for the pressure and density fluctuation relationship:

$$\text{p} = p_0 + \frac{\partial p}{\partial \rho}\big|_{\rho\,=\,\rho_0}(\rho - \rho_0) + \frac{1}{2}\frac{\partial^2 p}{\partial \rho^2}\big|_{\rho\,=\,\rho_0}(\rho - \rho_0)^2 + \ldots, \qquad (12.1)$$

where the partial derivatives are constant for the adiabatic process around the ρ_0 equilibrium density of the fluid.

If the density fluctuations are small, meaning $\bar{\rho} \ll \rho_0$, then the high-order terms can be reduced and the adiabatic state equation becomes linear:

$$\text{p} - p_0 = K\frac{\rho - \rho_0}{\rho_0}. \qquad (12.2)$$

The pressure generated by the sound p (12.4) is directly related with the particle movement and the amplitude ξ through equation (12.3):

$$\xi = \frac{v}{2\pi f} = \frac{v}{\omega} = \frac{p}{Z\omega} = \frac{p}{2\pi f Z}, \qquad (12.3)$$

$$\text{p} = \rho c 2\pi \text{f}\ \xi = \rho c \omega \xi = Z\omega\xi = 2\pi \text{f}\xi Z = \frac{aZ}{\omega} = Zv = c\sqrt{\rho E} = \sqrt{\frac{P_{ac}Z}{A}}, \qquad (12.4)$$

where the symbols together with the I.S. measurement units are presented in the following table:

The fundamental attenuation describes the power loss of a tone at a frequency f, during its movement across a distance d. The first level of our

Table 12.1 Symbols Definition and Corresponding I.S. Measurement Units

Symbol	Measurement Unit	Description
p	Pascal	Sound Pressure
f	Hertz	Frequency
ρ	kg/m^3	Environment Density (constant)
c	m/s	Sound Speed (constant)
v	m/s	Particle Speed
ω	rad / s	Angular Speed
ξ	m	Particle Movement
Z	N·s/m^3	Acoustic Impedance
a	m/s^2	Particle Acceleration
I	W/m^2	Sound Intensity
E	W·s/m^3	Sound Energy Density
P_{ac}	Watt	Acoustic Power
A	m^2	Surface

summary description takes into consideration this loss which occurs on the transmission distance *d*. The second level calculates the specific loss of one location caused by reflexions and refractions of upper and lower surfaces, i.e. sea surface and bottom and also, the sound speed variations due to depth differences. The result is a better prediction model of a specific transmitter. The third level addresses the apparently random power shifts of the signal received, by considering an average during a period of time. These changes are due to slow variations of the propagation environment, e.g. tidal waves.

All these phenomena are relevant for determining the transmission power needs in order to accomplish an efficient and successful underwater communication. We can also think at a separate model which could address much faster changes of the instantaneous signal power at any given time, but at a far smaller scale. The Signal Noise Ratio for different transmission distances as a frequency function can be viewed in Figure 12.8. The sound absorption limits the bandwidth which can be used for transmission and becomes dependent on the distance:

By evaluating the entity *A(d,f) N(f)* as a function of ideal propagation of the attenuation *A(d,f)* and as a consequence of tipical spectral power of the background noise *N(f)*, which drops 18dB per decade, we find the combined

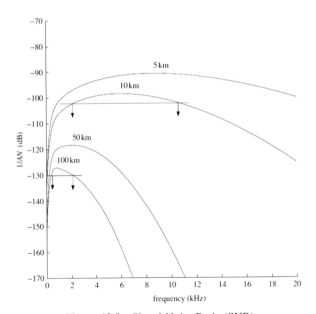

Figure 12.8   Signal-Noise Ratio (SNR).

effect of attenuation and noise in underwater acoustics. This characteristic describes the observation of SNR around the frequency bandwidth *f*. This shows that high frequencies suffer fast attenuation on long distances, which forces most modems to operate on a narrow bandwidth few kHz at most, by suggesting the optimal frequency for a specific transmission [26]. It also shows that the available bandwidth and implicitly the effective transmission rate reduces with higher distances; therefore, the development of any big network should start by determining its specific frequency and reserving a bandwidth around it [27].

12.2.3 Maintenance

Ocean can be a very harsh environment and underwater sensors are prone to failures because of fouling and corrosion. The sensor's construction method could include one miniaturized copper-alloy anode for anti-corrosion, as well as one miniaturized aluminum-alloy anode which could fight fouling. Modern anti-fouling systems already installed on rigs use microprocessor controlled anodes and the current flowing to each anti-fouling and anti-corrosion anode is quite low and the technology could be adapted by miniaturization of the existing anodes. Although we are considering the environmental impact of deploying such a high number of underwater devices, our primary concerns are the feasibility and the durability of the network and how we can address these factors in order to be able to expand our network through time and its enlargement to be backwards compatible to already deployed nodes. Besides the communication protocols being backwards compatible, underwater Safe-Net nodes must possess self-configuration capabilities, i.e. must be able to coordinate their operation, location or movement and data communication handshake protocols by themselves. So, we state the obvious, that this can only be possible if the Safe-Net nodes are resistant enough in the salt, corrosive water of the sea.

12.2.4 Law and Finance

At the end of 2010, the European Commission issued a statement concerning the safety regulations for offshore oil and gas drilling operations, with the declared purpose of developing new laws for the European Union concerning oil rigs. The primary objective of these measures will be the enforcement in this domain of the highest safety standards in the world until present time in order to prevent ecological disasters like the one in the Gulf of Mexico.

Moreover, during March – May 2011, following a public consultation session regarding the European Union legal frame for current practices of marine exploration and production safety standards, the community experts have drawn the line saying that although generally speaking the activities meet high standards of safety, these vary from one company to another, because of the different laws which apply in each country. Therefore, the next legislative proposal should enforce a common ground for all E.U. members concerning the laws for prevention, reaction times and measures in case of emergencies, as well as the *financial liability*.

According to the top 10 list of companies by largest revenues in the fiscal year 2010–2011, 7 are oil and gas industry companies, which summed up $2.54 billion USD revenues out of a total $3.43 billion USD. This means more than 74% of the global revenues [28]; therefore, the cost of deploying such Safe-Nets around drilling operations is rather small and the benefits would be huge. Laws could be issued by governments in order to enforce the obligation to oil and gas companies working at sea to use this sensor networks every time a new site is being surveyed or a new jacket is installed. This could also apply to existing oil rigs, jack-ups which move between different places or even for subsea production sites. The ability to have small devices physically distributed near offshore oil-fields' operations brings new opportunities for emergency-cases interoperability up to the higher level of disaster management systems point of view [29].

Ocean Sampling Networks have been experimented in the Monterey Bay area, where networks of sensors and AUVs, such as the Odyssey-class AUVs, performed synoptic, cooperative adaptive sampling of the 3D coastal ocean environment. Seaweb is an example of a large underwater sensor network developed for military purposes of detection and monitoring submarines [30]. Another example is the consortium formed by Massachusetts Institute of Technology (MIT) and Australia's Commonwealth Scientific and Industrial Research Organization which has collected data with fixed and mobile sensors mounted on autonomous underwater vehicles. The network was only temporary and lasted only for a few days around the coasts of Australia [31].

Ocean Observatories Initiative represents one of the largest ongoing underwater cabled networks, which has eliminated both the acoustic communication and power supply problems right from the start, by using already existing underwater cables or new installs. The investment on Neptune project was huge, approximately $153 billion dollars [32], but the idea seems quite bright if we look at the most important underwater cables, which are already running data under the oceans (Figure 12.9, with courtesy of TeleGeography.com). In 1956, North America was connected to Europe by an undersea cable called TAT-1. It was the world's first submarine telephone system, although telegraph cables had crossed for the ocean for a century. Trans-Atlantic cable capacity soared over the next 50 years, reaching a huge amount of data flowing back and forth the continents, nearly 10 Tbps in 2008.

12.2.5 Possible Applications

- Seismic monitoring. Frequent seismic monitoring is of importance in oil extraction; studies of variation in the reservoir over time are called 4-D seismic and are useful for judging field performance and motivating intervention;
- Disaster prevention and environmental monitoring. Sensor networks for seismic activity mentioned before could also be used for tsunami warnings to coastal areas. While there is always a potential for sudden devastation (see Japan 2011), warning systems can be quite effective. There is also the possibility of pollution monitoring: chemical, biological and so on and so forth;
- Weather forecast improvement: monitoring of ocean currents and winds can improve ocean weather forecasts, detecting climate change and also understanding and predicting the effect of human activities on marine ecosystems;
- Assisted navigation: sensors can be used to locate dangerous rocks in shallow waters. The buoys can also signal the presence of submerged wrecks or potential dangerous areas for navigation;
- Surveillance used for coast-line or border-lines, detecting the presence of ships in country marine belt. Fixed underwater sensors can monitor areas for surveillance, reconnaissance or even intrusion detection systems.

12.3 Mathematical Model

We introduce the class of systems which was considered when conducting the research for the PhD thesis, as well as definitions on configurations of sensors and remote actuators. This class of distributed parameter systems which describes important concepts for parameter identification and optimal experiment design has been adapted from the theoretical and practical research *"Optimal Sensing and Actuation Policies for Networked Mobile Agents in a Class of Cyber-Physical Systems"* [33]. The study presents models for

Figure 12.9 Most important underwater data & voice cables (2008).

aerial drones in the U.S.A., which could take high-resolution pictures of the agricultural terrain and the algorithm was pointed to data correlation between meteorological stations on the ground by matching the pictures with the low-resolution ones taken from satellites. The purpose was to introduce a new methodology to transform low-resolution remote sensing date about soil moisture to higher resolution information that contains better knowledge for use in hydrologic studies or water management decision making. The goal of the study was aiming to obtain a high-resolution data set with the help of a combination of ground measurements from instrumentation stations and low-altitude remote sensing, typically images obtained from a UAV. The study introduces optimal trajectories and launching points of UAV remote sensors in order to solve the problem of maximum terrain coverage using least hardware means, also expensive in their case.

We have taken further this study by matching the agricultural terrain with our underwater environment and making an analogy between the fixed instrumentation systems on ground, the meteorological stations and all the fixed offshore structures already put in place through-out the sea. The mobile drones are represented by remotely operated vehicles or by autonomous underwater vehicles which can have data collection sensors on-board and can be used as mobile network nodes. The optimisation of the best distribution pattern of the nodes in the underwater environment can be extrapolated only by neglecting the environment constants, which weren't taken into account by the study [33]. This issue is further to be investigated.

12.3.1 System Definition

The class of distributed parameter systems considered can be described by the state Equation [34]:

$$\begin{cases} \dot{y}(t) = Ay(t) + Bu(t) \\ \quad\quad y(0) = y_0 \end{cases} , \quad 0 < t < T, \tag{12.5}$$

where $Y = L^2(\Omega)$ is the state space and Ω is a bounded and open subset of $\mathbb{R}^n$ with a sufficiently regular boundary $\Gamma = \partial\Omega$. The domain Ω is the geometrical support of the considered system (12.5). A is a linear operator describing the dynamics of the as the set of linear maps from U to Y is the input operator; $u \in \mathcal{L}^2(0, T, U)$ is the space of integrable functions $f :]0, T[\mapsto U$ such that the function $t \rightarrow \|f(t)\|^p$ is integrable on $]0, T[$ and U is a Hilbert control space. In addition, the considered system has the following output equation:

$$z\left(t\right) = Cy\left(t\right), \tag{12.6}$$

Where $C \in \mathcal{L}\left(L^2\left(\Omega\right), Z\right)$ and Z is a Hilbert observation space. We can adapt the definitions of actuators, sensors, controllability and observability to system classes that are formulated in the state Equation form (12.5).

The tradition approach of the analysis in distributed parameter systems is fairly abstract in its purely mathematical form. Therefore, all the characteristics of the system related to its spatial variables and geometrical aspects of the inputs and outputs of the system are considered. To introduce a more practical approach from an engineering point of view, the study [33] introduces the concepts of actuators and sensors in the distributed parameter systems point of view. With these concepts at hand, we can describe more practically the relationship between a system and its environment, in our case sea/ocean water. The study can be extended beyond the operators A, B and C, with the consideration of the spatial distribution, location and number of sensors and actuators.

The sensors' measurements are, in fact, the observations on the system, having a passive role. On the other hand, actuators provide a forcing input on the system. Sensors and actuators can be of different natures: zone or point-wise or domain distributed, internal or boundary, stationary or mobile. An additional important notion is the concept of region of a domain. It is generally defined as a subdomain of Ω. Instead of considering a problem on the totality of Ω, the focus can be concentrated only on a subregion $\omega \varepsilon \Omega$, while the results can still be extended to $\omega = \Omega$. Such consideration allows the generalization of different definition and methodologies developed in previous works on distributed parameter systems analysis and control.

12.3.2 Actuator Definition

Let Ω be an open and bounded subset of $\mathbb{R}^n$ with a sufficiently smooth boundary $\Gamma = \partial\Omega$ [35]. An actuator is a couple (D, g) where D represents the geometrical support of the actuator, $D = supp\left(g\right) \subset \Omega$ and g is its spatial distribution.

An actuator (D, g) is said to be:

- A zone actuator if D is a non-empty sub-region of Ω;

- A point-wise actuator if D is reduced to a point $b \in \Omega$. In this case, we have $g = \partial_b$ where ∂_b is the Dirac function concentrated at b. The actuator is denoted (b, ∂_b).

An actuator, zone or point-wise, is said to be a boundary actuator if its support $D \subset \Gamma$. An illustration of the actuators supports is given in Figure 12.10:

In the previous definition, we assume that $g \in L^2(D)$. For a collection of p actuators $(D_i, g_i)_{1 \leq i \leq p}$, we have $U = \mathbb{R}^p$, $B : \mathbb{R}^p \to L^2(\Omega)$ and:

$$u(t) \to Bu(t) = \sum_{i=1}^{p} g_i u_i(t), \qquad (12.7)$$

$u = (u_1, u_2, \ldots, u_p)^T \in L^2(0, T, \mathbb{R}^p)$ and $g_i \in L^2(D_i)$ with $D_i = supp(g_i) \subset \Omega$ for $i = 1, \ldots, p$ and $D_i \cap D_j = \emptyset$ for $i \neq j$. So, we have the following:

$$B \cdot y = (g_1, y, g_2, y, \ldots, g_p, y)^T, \quad z \in L^2(\Omega), \qquad (12.8)$$

where M^T is the transpose matrix of M and $<\cdot, \cdot> = <\cdot, \cdot>_Y$ is the inner product in Y and for $v \in Y$, if $supp(v) = D$, we have:

$$\langle v, \cdot \rangle = \langle v, \cdot \rangle_{L^2(D)} . \qquad (12.9)$$

When D does not depend on time, the actuator (D, g) is said to be fixed or stationary. Otherwise, it is a moving or mobile actuator denoted by (D_t, g_t), where $D(t)$ and $g(t)$ are, respectively, the geometrical support and the spatial distribution of the actuator at time t, as in Figure 12.11:

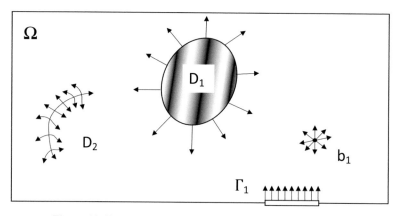

Figure 12.10   Graphical representation of actuators' supports.

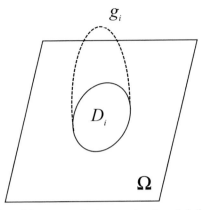

Figure 12.11   Illustration of the geometrical support and spatial distribution of an actuator.

12.3.3 Sensor Definition

A definition of sensors in the distributed parameters systems point of view is provided by [35]: a sensor is a couple *(D,h)* where D is the support of the sensor, $D = supp\,(h) \subset \Omega$ and h its spatial distribution.

A graphical representation of the sensors supports is given in Figure 12.12:

It is usually assumed that $h \in L^2\,(\mathrm{D})$. Similarly, we can define zone or point-wise, internal or boundary, fixed or moving sensors. If the output of the system is given by means of q zone sensors $(D_i, h_i)_{1 \le i \le q}$ with $h_i \in L^2\,(\mathrm{D}_i)$, $D_i = supp\,(h_i) \subset \Omega$ for $i = 1, \ldots, q$ and $D_i \cap D_j = \phi$ if $i \ne j$, then in the

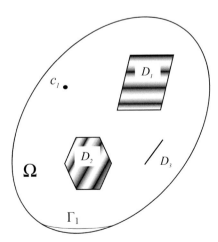

Figure 12.12   Graphical representation of the sensor supports.

zone case, the distributed parameter system's output operator C is defined by $C : L^2(\Omega) \to \mathbb{R}^p$:

$$y \to Cy = (h_1, y, h_2, y, \ldots, h_p, y)^T . \qquad (12.10)$$

And the output is given by:

$$z(t) = \begin{bmatrix} h_1, y_{L^2(D_1)} \\ \vdots \\ h_q, y_{L^2(D_q)} \end{bmatrix} . \qquad (12.11)$$

A sensor *(D,h)* is a zone sensor if D is a non-empty sub-region of Ω. The sensor *(D,h)* is a point-wise sensor if D is limited to a point $c \in \Omega$ and in this case $h = \partial_c$ is the Dirac function concentrated in c. The sensor is denoted as (c, ∂_c). If $D \subset \Gamma = \partial \Omega$, the sensor *(D,h)* is called a boundary sensor. If D is not dependent on time, the sensor *(D,h)* is said to be fixed or stationary, otherwise it is said to be mobile and is denoted as *(D_t, h_t)*. In the case of q point-wise fixed sensors located in $(c_i)_{1 \leq i = \leq q}$, the output function is a q-vector given by the relationship:

$$z(t) = \begin{bmatrix} y(t, c_1) \\ \vdots \\ y(t, c_q) \end{bmatrix} . \qquad (12.12)$$

Where c_i is the position of the i-th sensor and $y(t,c_i)$ is the state of the system in c_i at a given time t. [33] based on [36] devines also the notions of observability and local controllability in the sense of distributed parameters systems. [33] also shows that due to the nature of the problem of parameter identification, the abstract operator-theoretic formalism used above to define the dynamics of a distributed parameter system is not convenient. A formalism based on n partial differential equations is used instead. According to this setup, the sensor location and clustering phenomenon problem is ilustrated in the Fisher Information Matrix (FIM) [37], which is a well-known performance measure tool when looking for best measurements and is widely used in optimum experimental design theory for lumped systems [38]. Its inverse consititues an approximation of the covariance matrix for the estimate of Θ. However, there is a serious issue in the FIM framework of optimal measurements for parameter estimation of distributed parameters system, which is the dependence of the solution on the initial guess on parameters [39]. The dependence of the optimal location on Θ is very problematic; however, some *robust design* techniques have been developed in order to minimize or elude the influence and we propose similar methodologies.

By analogy with the study [33], we can try to optimize this solution for underwater points of interest. But in our case, of course, the problems are much more complex because of the physical and chemical properties of the environment.

We can consider two communication architectures for underwater Safe-Nets. One is a two-dimensional architecture, where sensors are anchored to the bottom of the ocean, and the other is a three-dimensional architecture, where sensors float at different ocean depths covering the entire monitored volume region. While the former is designed for networks whose main objective is to monitor the ocean bottom, the latter is more suitable to detect and observe phenomena in the three-dimensional space that cannot be adequately observed by means of ocean bottom sensor nodes. The mathematical model above refers only to the two-dimensional architecture case and we are looking into further researches for the three-dimensional optimization, especially when talking about the sensor-clustering phenomenon.

12.4 ROV

A remotely operated vehicle (ROV) is a non-autonomous underwater robot. They are commonly used in deep-water industries such as offshore hydrocarbon extraction. A ROV may sometimes be called a remotely operated underwater vehicle to distinguish it from remote control vehicles operating on land or in the air. ROVs are unoccupied, highly manoeuvrable and operated by a person aboard a vessel by means of commands sent through a tether.

They are linked to the ship by this tether (sometimes referred to as an umbilical cable), which is a group of cables that carry electrical power, video and data signals back and forth between the operator and the vehicle. The ROVs are used in offshore oilfield production sites, underwater pipelines inspection, welding operations, subsea BOP (Blow-Out Preventer) manipulation as well as other tasks:

- Seabed Mining – deposits of interest: gas hydrates, manganese nodules, metals and diamonds;
- Aggregates Industry – used to monitor the action and effectiveness of suction pipes during extraction;
- Cable and Node placements – 4D or time lapse Seismic investigation of crowded offshore oilfields;

- Jack-up & Semi-Submersible surveys – preparation and arrival of a Jack-up or Semi-Submersible drilling rig;
- Drilling & Wellhead support – monitoring drilling operations, installation/removal of template & Blow-Out Preventers (BOP), open-hole drilling (Bubble Watch), regular inspections on BOP, debris removal, IRM and in-field maintenance support (Well servicing);
- Decommissioning of Platforms / Subsea Structures – Dismantle structures safely and environment friendly;
- Geotechnical Investigation – Pipeline route surveys, Pipeline Lay – Startup, Touchdown monitoring, Pipeline Pull-ins, Pipeline crossings, Pipeline Lay-downs, Pipeline Metrology, Pipeline Connections, Post-lay Survey, Regular Inspections;
- Submarine Cables – Route Surveys, Cable Lay, Touchdown monitoring, Cable Post-Lay, Cable Burial;
- Ocean Research – Seabed sampling and surveys;
- Nuclear Industry – Inspections, Intervention and Decommissioning of Nuclear Power Plants;
- Commercial Salvage – Insurance investigation and assessment surveys, Salvage of Sunken Vessels, Cargoes, Equipment and Hazardous Cargo Recovery;
- Vessel and Port Inspections – Investigations, Monitoring of Ports and homeland security.

We are going to use PerrySlingsby Triton XLS and XLR models of the remote operated vehicles (ROV), which are currently available in the Black Sea area. While having the bigger goal in mind - deploying such networks on a large scale - we can only think now for a small test bed and before any physical implementation we are creating simulation scenarios on the VMAX ROV Simulator. Simulation helps preventing any damages to the ROV itself or any of the subsea structures we encounter. This also prevents any real-life impossible design-situations to occur, e.g.: the ROV's robotic arms have very good dexterity and their movement is described by many degrees of freedom - however, sometimes we find out the limits of motion and in some given situations, deploying objects in some positions may prove difficult or even impossible. We address these hypothetical situations and try to find the best solutions for securely deploying the sensors by anchors to the sea floor or by tethering to any metallic or concrete structures: jackets, jack-up legs, autonomous buoys, subsea well production heads, offshore wind farm production sites, so on and so forth.

In the Black Sea area, operating in Romania's territorial sea coast line, we identified 4 working-class ROVs, out of which 2 are manufactured by PerrySlingsby U.K.: 1 Triton XLX and 1 Triton XLR - first prototype of its kind, which led to our models used in simulation.

12.4.1 ROV Manipulator Systems

Schilling Robotics' TITAN 4 manipulator with 7 degrees of freedom (Figure 12.13) is the industry's premier system that offers the optimum combination of dexterity and strength. Hundreds of TITAN manipulators are in use worldwide every day, and are the predominant manipulator of choice for work-class ROV systems. Constructed from titanium, the TITAN 4 is uniquely capable of withstanding the industry's harsh environment and repetitive needs.

The movement of the 7-function Master Arm Control (Figure 12.14) is transmitted through the fibre optics inside the tether and the underwater media converters situated on the ROV pass the information to the Titan 4 Manipulator after it is checked for send/receive errors. The exact movement of the joints of the 7-function above the sea level represent the movement of the Titan-4

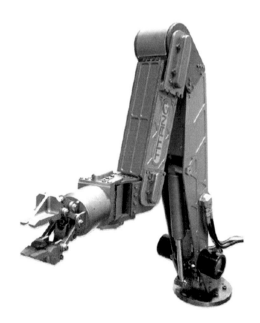

Figure 12.13   Titan 4 Manipulator 7-F.

Figure 12.14   Master Arm Control.

underwater. This provides the dexterity and degrees of freedom needed to execute most difficult tasks (Figure 12.15):

Schilling's RigMaster is a five-function, rate-controlled, heavy-lift grabber arm that can be mounted on a wide range of ROVs (Figure 12.16). The grabber arm can be used to grasp and lift heavy objects or to anchor the ROV by clamping the gripper around a structural member at the work site.

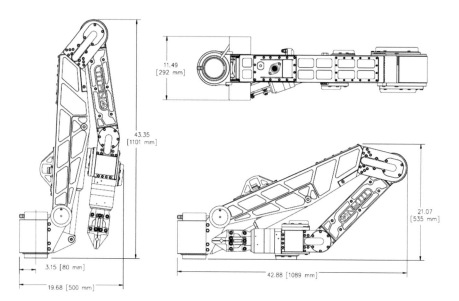

Figure 12.15   Titan 4 – Stow dimensions.

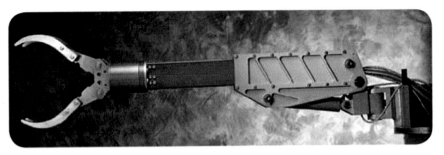

Figure 12.16   RigMaster 5-F.

Constructed primarily of aluminium and titanium, the RigMaster delivers the power, performance, and reliability required for such demanding work. A typical work-class ROV utilizes a combination of the five-function RigMaster and seven-function TITAN 4.

With these two manipulator systems, any type of sensor can be deployed or fixed on the ocean bottom. In order for a better understanding of the process and likely problems which can occur during the installation, we are going to use the VMAX Tech. – PerrySlingsby ROV Simulator for which we are going to develop a modelling and simulation scenario concerning the deployment of

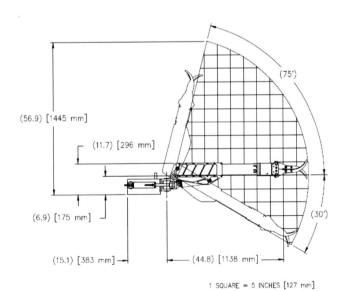

Figure 12.17   RigMaster range of motion – Side view.

underwater sensors Safe-Net surrounding areas of offshore oil and gas drilling operations.

12.4.2 Types of Offshore Constructions

Offshore constructions represent the installation of structures and facilities in a marine environment, usually for the production and transmission of electricity, oil, gas or other resources. We have taken into consideration most usual encountered offshore types of structures and facilities, focusing on the shapes which are found underwater:

- Fixed platforms;
- Jack-up oil and gas drilling and/or production rigs;
- Jackets with top sides;
- Spars or floating platforms;
- Semi-submersibles;
- Drilling ships;
- Floating tension legs;
- Floating production storage and offloading (FPSO);
- Subsea well production heads;
- Offshore wind farm production sites.

We have created a simple scenario in which we use a PerrySlingsby Triton XLS ROV connected to a TMS (Tether Management System) and where we can use the physics of the robotic arms in order to understand which movements are going to be needed in order to implant sensors of different sizes into the ocean floor, as well as nearby the types of subsea structures mentioned above. We try to create handling tools for the Schilling Robotics 7-F arm in order to easily deploy and fix or common framework device model and also try to find best spots for all the offshore types of structures we encountered in our offshore experience inquiry [40].

12.5 ROV Simulator

The VMAX Simulator is software and hardware package intended to be used by engineers to help in the design process of procedures, equipment and methodologies, having a "physics based simulation" for the offshore environment. The simulator is capable of creating scenarios that are highly detailed and focused on one area of operation or broad in scope to allow an inspection of an entire subsea field. For creating a scenario, there are two skill sets needed: 3D Studio Max modelling and ".lua" script programming skills.

In order to safely deploy our Safe-Nets' sensor balls into the water and fix them to jack-up rigs metallic structures or to any other offshore constructions, we first try to develop models of those structures and include them into a standard fly-alone ROV simulation scenario. This is a two-step process as any object's model has to be created in 3D Studio Max software and afterwards it can be programmatically be inserted into the simulation scenario. The simulation scenarios are initialized by a series of Lua scripts, which is very similar to C++ programming language and The VMAX Scenario Creation is *open source*. The scripts are plain text files that can be edited using many programs, including Microsoft Windows Notepad. The file names end with .lua extension and are recommended to be opened with jEdit editor. This is also an open-source editor which requires the installation of Java Runtime Environment (JRE).

We have altered the simulation scenarios as it can be seen in Figure 12.18 and Figure 12.19 in order to obtain a better model of the Black Sea floor through-out Romania's coast line, which usually contains more sand because of the Danube sediments coming from The Danube Delta. Geologists working on-board the Romanian jack-ups considered the sea-floor in the VMAX ROV Simulator very much alike with the one in the geological and oil-petroleum interest zones up to 150-160 miles out in the sea. Throughout these zones the water depth doesn't exceed 80-90m, which is the limit at which drilling jack-up rigs can operate (legs have 118m in length).

The simulator which is open-source was the starting base for a scenario where we translated the needs of the ROV in terms of sensor handling, tether positioning and pilot techniques combined with the specifications of the sea-floor where the Safe-Nets will be deployed. The scenarios are initialized by a series of .Lua scripts and the typical hierarchical file layout is presented in Figure 12.20.

The resources are divided into two large classes of information: *Scenarios*-related data and *Assets*. The former contains among others: Bathymetry, Lua, Manipulators, Tooling, TMS (Tether Management System), Vehicles, Components and IP (Internet Protocol communications between assets).

Bathymetry directory contains terrain information about a specific location, where we could alter the sand properties on the sea floor. The terrain stored here may be used across several scenarios. We could add a graphic asset by using the template for the bathymetry part. The collision geometry can be later generated based on the modelled geometry. We remind that the

Figure 12.18   Triton XLS ROV in simulation scenario.

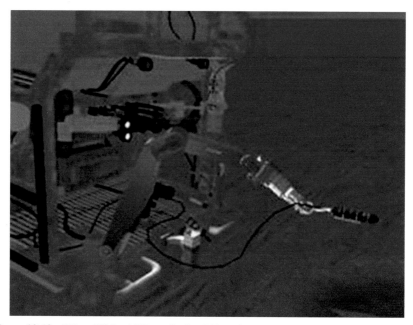

Figure 12.19   Triton XLS schilling robotics 7-Function arm in scenario.
Courtesy of TelegeoGraphy.com

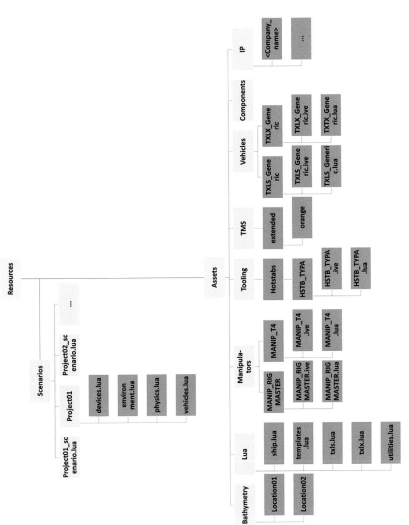

Figure 12.20 Typical hierarchical file layout.

simulator scenario creation software is *open-source* and we present in the following lines some of the parts of the basic scenario provided with the full-up simulator which we modified in order to accommodate our specific needs:

```
graphicAsset {
assetID = "bathymetry",
castShadow = true,
– can be false for very flat terrain
– Our terrain model =
"assets/Bathymetry/TER_500m_v2.0/TERBLKSEA_500m_v1.0.ive",
receiveShadow = true,
scale = { 2, 2, 2 }
– specific to this particular model
}
–We changed the environment table to look like this:
environment = {
assembly = {
– Various items in the environment starting with bathymetry.
parts = {
– add the bathymetry based on a template
createFromTemplate(templates.bathymetry,
{
collisions = {
– The first item in the array is for the collision
– geometry automatically created from the model.
{
– set the area over which the bathymetry spans
size = { 100, 100, 1 },
– must be specified
}
– collision primitives may be appended to this array
},
– set the depth of the bathymetry
position = { 0, 0, REFERENCE_DEPTH - 20 }}),}
constraints = { },
selfCollide = true,},
bathymetryPartName = "bathymetry",
pickFilter = { "bathymetry" },
```

```
currentDirectionTable = { 0 },
currentSpeedTable = { 1 },
depthTable = { 0 }}
```

The bathymetry template uses triangle mesh collision for the terrain. This will provide collisions that are contoured to the bathymetry model

The Manipulators directory contains sub-directories for each arm and each sub-director contains a model file with a .Lua script function used to create the manipulator and add it to the ROV. We are looking forward to creating a new manipulator usable for each case of sensor deployment.

The Tooling directory has either categories of tools or some uncategorized ones, each having a model file name ".ive" or ".flt" and a Lua script file with the code to create that specific tool [41].

Whereas the typical training scenarios include mainly a fly-around and getting used to the ROV commands for the pilot and assistant pilot, we have used the auxiliary commands in order to simulate the planting of the Safe-Net around a jacket or a buoy for example. As far as the training scenario is concerned, we covered the basics for a pilot to get around a jacket safely, carrying some sort of object in the Titan4 manipulator robotic arm, without dropping it, or without having the ROV's tether tangling with the jacket metallic structure. The tether contains hundreds of fibre-optic cables covered with a Kevlar reinforcement, but it is recommended that no more than 4 total 3600 spins are made in one direction, clockwise or counter-clockwise, even having this strengthened cover, in order to avoid any loss of communication between the control console and the ROV itself. Any interaction between the tether and the metallic structure could represent a potential threat to the ROV's integrity.

12.6 Common Modular Framework

An overview of the challenges and application possibilities of deploying underwater sensor networks nearby oil rigs drilling operations areas and offshore construction sites surroundings led to the conclusion that a standard device is needed in order to deploy multi-purpose underwater sensor networks. We detected the need for a standard, common, easy-to-use device framework for multi-purpose underwater sensors in marine operations, as we were preparing the devices for future use. This framework should be used for multiple different sensors and we consider the modular approach to be best suited for future use, providing the much-needed versatility.

We considered the buoyancy capabilities needed for a stand-alone device launched on the sea surface and we started with an almost spherical-shaped model Figure 12.21. If tethering should be needed, a small O-ring cap on one of the sphere's poles can be mounted:

The device will be able to accommodate a variety of sensors, adapted within the inside "drawers", its layers being highly modular. In this manner, with the same network node, we will be able to empower a lot of types of applications and this is an essential step in justifying the costs of development. We believe this characteristic is critical for improving the financial desirability of any future Safe-Nets offshore project.

Our simulation scenario is still scarce in modelled objects as the process of creating them quite realistic is taking a long time. However many simulation scenario variables we may alter, after finding out real types of situations which occur on offshore structures, we learned that simulating the deployment and deciding spots of anchoring for our sensors can only help, but not solve real-life deployment, as parameters decided beforehand on shore can change dramatically offshore.

However, we believe that our 3D models for underwater multi-purpose sensors still stand as a good idea for our Safe-Nets real-life development and implementation. Tethered or untethered, the upper hemisphere can include a power adapter which can be used also as batteries compartment if the sensor is wireless. The sensors have enough drawers for electronic modules and Type 03 is designed with built-in cable management system. Also, Type 03 is designed with a membrane for a sensitive pollution sensor. We have chosen a very simple closing mechanism for starters, using clamps on both sides, which can

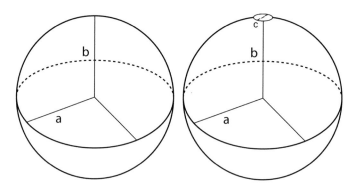

Figure 12.21 Spherical-shaped model designed for common framework; a $>=$ b; c is Tether/Cable entry point diameter.

ensure the sealing of the device. The upper and lower hemispheres close on top of an O-ring seal which can be lubricated additionally with water repellent grease. Also, we have designed a unidirectional valve which can be used for a vacuum pump to clear out the air inside. The vacuum strengthens the seal against water pressure. In Figure 12.22, we present a few prototypes which we tried to model and simulate:

Within the same common modular framework, we have thought at a multi-deployment method for 3 or more sensors at the same time. Actually, the following ideas were issued because of the repeated fail trials with an ROV to grab and hold a Safe-Net sensor long enough in order to place it in a hook coming from an autonomous buoy above the sea surface, affected by wave length and height. Because of the real difficulties encountered, especially when inserting higher waves into the scenario, we have thought of a way to get the job done more quickly (Figure 12.23):

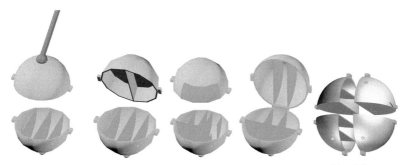

Figure 12.22   Underwater multi-purpose devices prototypes 01 – 05.

Figure 12.23   Grouping method for multiple simultaneous sensor deployment.

Moreover, the spherical model framework of the sensor, the basic node of the Safe-Net, will prove to be very difficult to handle using the simple manipulator, as it tends to slip, and the objective is to carry it without dropping. Therefore, we have designed a "cup-holder" shape for grabbing more easily the sphere and if it contains a cable connection, it should not be tampered by the grabber, as it can be seen in Figure 12.24:

12.7 Conclusions

Most of the state-of-the-art solutions regarding underwater sensor networks rely on specific task-oriented sensors, which are developed and launched with different means and no standardization. The studies we found usually use power from batteries and all sorts of resilient algorithms in order to minimize battery draining and use sleep-awake states of the nodes, which finally are recovered from water in order to retrieve data collections. Our approach is trying to regulate the ways of deploying and fixing the sensors towards offshore structures and moreover to offer solutions to more than one application task. This may seem as a general approach, but this is needed in order to avoid launching different technology nodes which afterwards will not be able to communicate with each other. Development of a virtual environment-based training system for ROV pilots could be the starting point for deploying underwater sensor networks worldwide, as these are the people who will actually be in the position to implement it.

This chapter investigates the main challenges for the development of an efficient common framework for multi-purpose underwater data collection

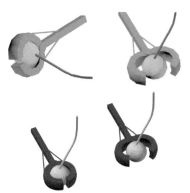

Figure 12.24   Basic sensor device holder designed for simulation.

devices (sensors). Several fundamental key aspects of underwater acoustic communications are also investigated. We want to deploy around existing offshore constructions and this research is still a work in progress. The main challenges for the development of efficient networking solutions posed by deploying sensors in the underwater environment are detailed at all levels. In short, this article has analyzed the necessity of considering the physical fundamentals of an underwater network development in the planetarium ocean, starting from the instrumentation needs surrounding offshore oil drilling sites and early warning systems for disaster prevention worldwide.

We suggest various extension possibilities for applications of these safenets, starting from pollution monitoring around offshore oil drilling sites, early warning systems for disaster prevention (earthquakes, tsunami) or weather forecast improvement, up to military surveillance applications, all in order to overcome the cost of implementation of such underwater networks.

References

[1] K. Eshghi and R. C. Larson, 'Disasters: lessons from the past 105 years', Disaster Prevention and Management, Vol. 17, pp.61–82, 2008.

[2] D. Green, 'Acoustic modems, navigation aids and networks for undersea operations', IEEE Oceans Conference Proceedings, pp.1–6, Sydney, Australia, May 2010.

[3] J. Heidemann, Y. Li and A. Syed, 'Underwater Sensor Networking: Research Challenges and Potential Applications', USC Information Sciences Institute, USC/ISI Technical Report ISI-TR-2005–603, 2005.

[4] T. Melodia, Ian F. Akyildiz and D. Pompili, 'Challenges for Efficient Communication in Underwater Acoustic Sensor Networks', ACM Sigbed Review, vol.1, no.2, 2004.

[5] A. Cerpa and et al., 'Habitat monitoring: Application driver for wireless communications technology', ACM SIGCOMM Workshop on Data Communications in Latin America and the Caribbean, San Jose, Costa Rica, 2001.

[6] D. Whang, N. Xu and S. Rangwala, 'Development of an embedded sensing system for structural health monitoring', International Workshop on Smart Materials and Structures Technology, pp. 68–71, 2004.

[7] D. Steere, A. Baptista and D. McNamee, 'Research challenges in environmental observation and forecasting systems', 6th ACM International Conference on Mobile Computing and Networking, Boston, MA, USA, 2000.

[8] L. Dignan, 'Google's Data Centers', [Online] 2011.http://www.zdnet.com /blog/btl/google-makes-waves-and-may-have-solved-the-data-center-conundrum/9937http://www.datacenterknowledge.com/archives/2008/ 09/06/google-planning-offshore-data-barges/.

[9] A. S. Outlaw, 'Computerization of an Autonomous Mobile Buoy', Florida Institute of Technology, Vol. Master Thesis in Ocean Engineering, Melbourne, FL, 2007.

[10] C. Garcier, et al., 'Autonomous Meteorogical Buoy', Instrumentation Viewpoint, vol. 7, Winter, 2009.

[11] V. Dache, M.C. Caraivan and V. Sgarciu, 'Advanced Building Energy Management Systems - Optimize power consumption', INCOM 2012, pp. 426, Bucharest, 2012.

[12] C. Hong-Jun, et al., 'Challenges: Building Scalable and Distributed Underwater Wireless Sensor Networks (UWSNs) for Aquatic Applications', UCONN CSE Technical Report, UbiNet-TR05–02, 2005.

[13] D. Pompili, T. Melodia and A.F. Ian, 'A Resilient Routing Algorithm for Long-term Applications in Underwater Sensor Networks', Atlanta, 2005.

[14] Aquaret, [Online], 2008. www.aquaret.com

[15] PelamisWaves, [Online], 2012. http://www.pelamiswave.com/pelamis-technology

[16] WaveBob, [Online], 2009. http://www.wavebob.com

[17] Buoy, OE, [Online], 2009. www.oceanenergy.ie

[18] PowerBuoy, [Online], 2008. www.oceanpowertechnologies.com

[19] Oyster, [Online], 2011. www.aquamarinepower.com

[20] WaveDragon, [Online], 2011. www.wavedragon.net

[21] AWS, [Online], 2010. http://www.awsocean.com

[22] F. Mosca, G. Matte and T. Shimura, 'Low-frequency source for very long-range underwater communication', Journal of Acoustical Society of America Express Letters, vol. 133, 10.1121/1.4773199, Melville, NY, U.S.A., 20 December 2012.

[23] D. Pompili and T. Melodia, 'An Architecture for Ocean Bottom UnderWater Acoustic Sensor Networks (UWASN)', Georgia, Atlanta, 2006.

[24] R. Urick, 'Principles of underwater sound', McGraw Hill Publishing, New York, NY, U.S.A., 1983.

[25] S.W. Rienstra and A. Hirschber, 'An Introduction to Acoustics', Eindhoven University of Technology, Eindhoven, The Netherlands, 2013.

[26] J. Wills, W. Ye and J. Heidemann, 'LowPower Acoustic Modem for Dense Underwater Sensor Networks', USC Information Sciences Institute, 2008

[27] M. Stojanovic, 'On the relationship between capacity and distance in an underwater acoustic communication channel', ACM Mobile Computing Communications, Rev.11, pp.34–43, doi:10.1145/1347364.1347373, 2007.

[28] Wikipedia.org, Wikipedia List of Companies by Revenue, [Online], 2011. http://en.wikipedia.org/wiki/List_of_companies_by_revenue

[29] V. Nicolescu, M. Caraivan, 'On the Interoperability in Marine Pollution', IESA'14 7th International Conference on Interoperability for Enterprise Systems and Applications, Albi, France, 2014.

[30] J. Proakis, J. Rice, et al., 'Shallow water acoustic networks', IEEE Communications Magazine, pp. 114–119, 2001.

[31] I. Vasilescu, et al., 'Data collection, storage and retrieval with an underwater sensor network', 3rd ACM SenSys Conference Proceedings, pp.154–165, San Diego, CA, U.S.A., November 2005.

[32] P. Fairley, 'Neptune rising', IEEE Spectrum Magazine #42, pp. 38–45, doi:10.1109/MSPEC.2005.1526903, 2005.

[33] C. Tricaud, 'Optimal Sensing and Actuation Policies for Networked Mobile Agents in a Class of Cyber-Physical Systems', Utah State University, Logan, Utah, 2010.

[34] A. El Jai, 'Distributed systems analysis via sensors and actuators', Sensors and Actuators A', vol. 29, pp.1–11, 1991.

[35] A. El Jai and A.J. Pritchard, 'Sensors and actuators in distributed systems', International Journal of Control, vol. 46, iss. 4, pp. 1139–1153, 1987.

[36] A. El Jai, et al., 'Regional controllability of distributed-parameter systems', International Journal of Control, vol. 62, iss. 6, pp.1351–1356, 1995.

[37] E. Rafajowics, 'Optimum choice of moving sensor trajectories for distributed parameter system identification', International Journal of Control, vol. 43. pp.1441–1451, 1986.

[38] N.Z. Sun, 'Inverse Problems in Groundwater Modeling', Theory and Applications of Transport in Porous Media, Kluwer Academic Publishers, Dodrecht, The Netherlands, 1994.

[39] M. Patan, 'Optimal Observation Strategies for Parameter Estimation of Distributed Systems', University of Zielona Gora Press, Zielona Gora, Poland, 2004.

[40] M. Caraivan, V. Dache and V. Sgarciu, 'Simulation Scenarios for Deploying Underwater Safe-Net Sensor Networks Using Remote Operated Vehicles', 19th International Conference on Control Systems and Computer Science Conference Proceedings, IEEE CSCS'19 BMS# CFP1372U-CDR, ISBN: 978–0-7695–4980-4, Bucharest, Romania, 2013.

[41] B. Manavi, VMAX Technologies Inc. Help File, Houston, 77041–4014 Texas, TX, U.S.A., 2010.

13

M2M in Agriculture – Business Models and Security Issues

S. Gansemer[1], J. Sell[1], U. Grossmann[1], E. Eren[1], B. Horster[2], T. Horster-Möller[2] and C. Rusch[3]

[1]University of Applied Sciences and Arts Dortmund, Dortmund, Germany
[2]VIVAI Software AG, Dortmund, Germany
[3]Claas Selbstfahrende Erntemaschinen GmbH, Harsewinkel, Germany
Corresponding author: S. Gansemer <sebastian.gansemer@fh-dortmund.de>

Abstract

Machine-to-machine communication (M2M) is one of the major innovations in the ICT sector. Especially in agricultural business with heterogeneous machinery, diverse process partners and high machine operating costs, M2M offers large potential in process optimization. Within this paper, a concept for process optimization in agricultural business using M2M technologies is presented using three application scenarios. Within that concept, standardization and communication as well as security aspects are discussed. Furthermore, corresponding business models building on the presented scenarios are discussed and results from economic analysis are presented.

Keywords: M2M, agriculture, communication, standardization, business case, process transparency, operation data acquisition, business model, security.

13.1 Introduction

Machine-to-machine communication (M2M) currently is one of the major innovations in the ICT sector. The agricultural sector is characterized by heterogeneous machinery, diverse process partners and high operational

Advances in Intelligent Robotics and Collaborative Automation, 295–312.

machinery costs. Many optimization solutions aim to optimize a single machine but not the whole process. This paper deals with improving the entire process chain within the agricultural area. In the first part of this paper, a concept for supporting process optimization in heterogeneous process chains in agricultural business using M2M communication technologies is discussed. The second part presents business cases for the proposed system and outcomes from economic analysis. In the third part last not least security aspects related to the proposed system are discussed.

13.2 Related Work

The application of M2M technology in agriculture is targeted by several other research groups. Moummadi et. al. [1] present a model for an agricultural decision support system using both multi-agent-system and constraint programming. The systems purpose is controlling and optimizing water exploitation in greenhouses.

Wu et. al. [2] present a number of models for M2M usage in different sectors such as utilities, security and public safety, tracking and tracing, telematics, payment, healthcare, remote maintenance and control and consumer devices. They discuss technological market trends and the influence of different industries on M2M applications.

An insurance system based on telematics technology is demonstrated by Daesub et. al. [3]. They investigate trends in insurance industry based on telematics and recommend a supporting framework.

A business model framework for M2M business models based on cloud computing is shown by Juliandri et. al. [4]. They identify nine basic building blocks for a business model aiming to increase value while reducing costs.

Gonçalves and Dobbelaere [5] discuss several business scenarios based on specific technical scenarios. Within the presented scenarios, the stakeholders assume different levels of control over the customer relationship and the assets determining the value proposition.

A model for software updates of mobile M2M devices is presented in [6]. They aim on low bandwidth use and avoidance of system reboot.

13.3 Communication and Standardization

The agricultural sector is characterized by heterogeneous machinery and diverse process partners. Problems arise from idle times in agricultural processes, suboptimal machine allocation and improper planning. Other problems are generated by incompatibilities of machinery built by different manufactur-ers. Because of proprietary signals on machine buses not fitting on one another collaboration between machines may be inhibited [7, 8].

To support collaboration of heterogeneous machinery a standardized communication language is needed. Communication takes place either direct via machine to machine or via machine to cloud.

Sensors in machines record different parameters such as position, moving speed, mass and quality of harvested produce. These operational and machine logging data from the registered machines are synchronized between machines and finally sent via telecommunication network to a recording web portal. Data are stored within the portal´s database and are used for optimizing process chain or develop and implement business models based on that data. All data is sent through machine's ISO- and CAN-bus in proprietary syntax.

Within the concept, each machine uses a "black-box" which trans-lates manufacturer specific bus signal data to a standardized data format. The concept is shown in Figure 13.1. Machines may be equipped with diverse numbers of sensors resulting in different numbers of signals available. The standard should cover most of those signals. However, due to the diverse machinery available, not every signal available on the machine can be supported within the proposed concept.

Within this paper, the concept of a portal (M2M-Teledesk) is presented suited for dealing with the problems mentioned above. The system's framework is shown in Figure 13.2. The black-boxes installed on each machine are interfaces between the machine's internal buses and the portal

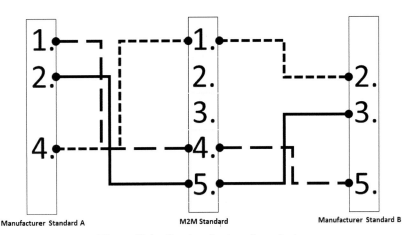

Figure 13.1   Synchronization of standards.

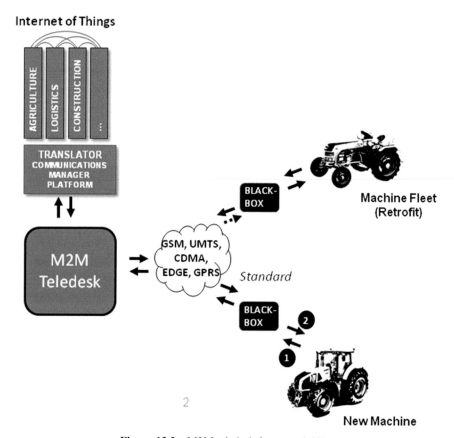

Figure 13.2 M2M teledesk framework [9].

(M2M-Teledesk). Black-boxes are equipped with mobile network communication interfaces for transferring data between machines among each other and between machines and the portal as well.

Every machine is set up with a black-box which reads internal buses and translates signals to the proposed open M2M standard, runs different applications and communicates data to and from the machine using WIFI or mobile data communication networks. The system uses a public key infrastructure for safety and trust reasons (see Section 13.7). Within the portal collected data is aggregated and provided to other analyzing and evaluating systems (e.g. farm management). Depending on the machine a full set or a subset of data specified in the standard can be used. Older machines may be retrofitted with a black-box providing only a subset of available data as a smaller number of sensors are available only.

The data is visualized within the portal and helps the farmer to optimize business processes to meet documentation requirements or to build data-based business models. Especially when it comes to complex and detailed records of many synchronized machines, the system shows its advantages.

Communication between machines takes place either directly from machine to machine or via a mobile communication network (e.g. GSM or UMTS). Within agricultural processes operating in rural areas, the availability of mobile communication networks is not always given. There are two strategies to increase the availability of network coverage:

- National roaming SIM cards;
- Femtocells.

With national roaming SIM cards being able to roam into all available networks, the availability of mobile network coverage can be increased, while with standard SIM cards only one network can be used in the home country [10]. National roaming SIM cards are operating in a country different from their home location (e.g. a spanish SIM card operating in Germany). The SIM card can roam into all available networks as long as issuing provider and network operator signed a roaming agreement. Although network coverage can be increased, a communication channel cannot be guaranteed.

With femtocells [2], dedicated base station is placed on the field where machines are operating. The concept is presented in Figure 13.3. Machines

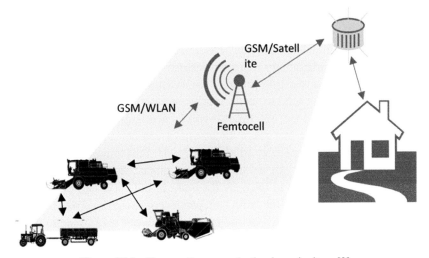

Figure 13.3 Femtocell communication in agriculture [9].

communicate to the base station e.g. via WLAN or GSM/UMTS, while base-station is connected to the portal by GSM/UMTS or satellite connection. The location of the femtocell base-station should be chosen in a way that coverage is given at every location within the corresponding area either via the femtocell or via direct connection to a mobile network. This strategy enables communication even without network coverage by the operator. However, the implementation effort is significantly higher than in case of using national roaming SIM cards.

13.4 Business Cases

The described system can be used in different manners. Three main business cases have been identified:

- Process Transparency (PT);
- Operation Data Acquisition (ODA);
- Remote Software Update (RSU).

Process transparency (PT) mainly focuses on in-time optimization of process chains, while ODA uses downstream analysis of data. Remote software update (RSU) aims to securely install applications or firmware updates on machines without the use of a service technician. These three business cases are described below in more detail.

13.4.1 Process Transparency (PT)

Processes in agricultural business are affected by several process participants. Furthermore, the used machines in many cases are operating with high costs. A visualization of an exemplary corn-harvesting process is presented in Figure 13.4. During the harvesting process, a harvester is e.g. cropping corn. Synchronously, a transport vehicle needs to drive in parallel to the harvester to transport the harvested produce. Machines involved in this sub-process need to be synchronized in real time. In case of the transport vehicle being filled up, it has to be replaced by another empty transport vehicle. Full transport vehicles make their way to e.g. a silo or a biogas power plant where the transport vehicle has to enter via a scale to measure the mass of the harvested produce. Furthermore, a quality check of the harvested produce is carried out manually.

   This process may be optimized by the portal in different ways. Due to the registration of sensor data, the weighting and quality check part in the process may be skipped or reduced to spot checks if the customer deems the data within the system to be trustworthy. Furthermore, the data is visualized by the

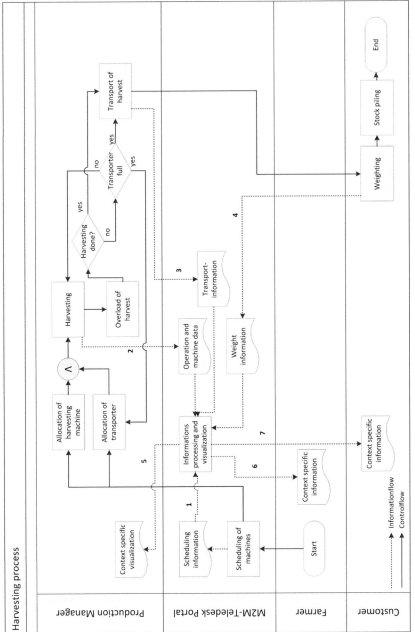

Figure 13.4 Information- and controlflow of scenario harvesting process.

portal to give the production manager the opportunity to optimize the process in near real-time. Before starting the process, a production plan is prepared by the production manager either manually with support by the system or automatically by the system. Within the plan, machines are allocated with time and position data. When the system registers a plan deviation, the plan is updated either manually or automatically. This approach allows reducing idle times saving costs and resources.

13.4.2 Operations Data Acquisition (ODA)

Within Operations Data Acquisition (ODA) scenario data gathered by the machine sensors is saved for downstream processing and analysis. While process transparency aims to synchronize process data in real-time to support process optimization, ODA data is gathered and sent to the server after the process is finished. Analysis is done e.g. to generate yield maps or to analyze machine behavior.

13.4.3 Remote Software Update (RSU)

The remote software update (RSU) process aims to remotely install software on a machine. Software update includes two sub scenarios, firmware upgrade and app-installation. App-installation means the installation of an additional piece of software from a third-party-software provider while firmware updates. The main aspect of software update is to ensure that the software is installed in a secure way, meaning that the machine proof to install software which comes from an authorized source and was not changed during network transport. Details on the security measures can be found in Section 13.7.

13.5 Business Models

Based on the scenarios and the data described above business and licensing models are developed. Figure 13.5 shows the value chain of M2M Teledesk consisting of six partners.

For all partners of the value chain business potential has been analyzed and is shown in Table 13.1. The table shows the partner's roles, the expected revenue and cost development and the resulting business potential.

Figure 13.5   Value chain of M2M-Teledesk [9].

Table 13.1　Revenue, costs and business potential for partners along M2M value chain

Partner	Role	Revenue Development (Per Unit)	Cost Development (Per Unit)	Business Potential
Module manufacturer	Manufacturer of black-box	Constant	Declining	+
Machine manufacturer	Manufacturer of machines	Progressive	Declining	++
Mobile network operator	Data transport, SIM management	Constant	Declining	+
3rd party software provider	Software developer, application provider	Constant/ progressive (depending on business model)	Depending on business model	+
Portal provider	Portal operator	Progressive	Declining	++

The module manufacturer produces the black-boxes (see Figure 13.2) built into the machines or used to retrofit older machines. Revenues for module manufacturers mostly come from black-box sales. Costs per unit are expected to decline with increasing number of sold units.

The machine manufacturer's revenues come from machine sales as well as services delivery and savings due to remote software updates. The cost of development is expected to be declining with the increasing number of sold units.

The mobile network operator's role is to deliver data through a mobile network. SIM card management may also be done by the network operator but can also be done by an independent partner. Revenues consist of fees for data traffic as well as service fees for SIM card supply and management. Additional costs for extra data volume over an existing network are very low.

Third-party software providers can be part of the value chain; however, this is not compulsory. They either supply an application bringing additional functions to the machinery or implement an own business model based on the data held in the portal.

The software is sold through the portal and is delivered to the machinery by the remote software update process described above. The revenues development per unit depends on the employed business model. When only software is sold, revenues per unit are constant. With additional business models, revenues may also develop progressively.

Costs are mostly one-time costs for software development as well as running costs for maintenance. However, additional costs may arise depending on the business model. The portal provider operates and manages the portal. Revenues consist of usage fees, revenues from third party app sales, fees for delivering software updates and other service fees. Costs are mainly for portal operation, support and data license fees. The end users' revenues come from savings due to increased process efficiency, while costs arise for additional deductions for machines, additional costs for higher skilled workforce, system usage fees and so on. Business potential is given for all partners involved in the value chain.

With applications developed by third-party software developers a variety of new business models can be implemented. One model is given by "pay-per-use" as well as "pay-how-you-use" insurance or leasing. Within this business model insurance or leasing companies are able to calculate insurance or leasing rates more adequate to risk depending on real-usage patterns. The insurance or leasing company is integrated in the value chain as a third-party software provider. For running the business model, data showing the usage pattern is needed. To gain this data, the third-party software provider needs to pay license fees.

13.6 Economic Analysis

Economic analysis of the system leads to a model consisting of linear equations. For visualizing the quantitative relations between different services, sub-services and partners of a so-called swimlane-gozintograph is used. Based on standard gozintograph methodology as described in [11] the resulting figure is adapted by including swimlane methodology [12] to show the involved partners. Figure 13.6 shows the corresponding swimlane-gozintograph. Columns represent the involved partners; transparent circles indicate different services delivered by the partners. Shaded circles represent business cases, i.e. services delivered externally.

The figure shows the relations between internal and external services and the share of each partner in the different business cases. From this gozintograph, mathematical equations can be derived, enabling the calculation of the gross margins for each business case.

From Figure 13.6, linear equations are derived, including transfer prices, amounts of service delivery and external sales prices for cost and gross margin calculation.

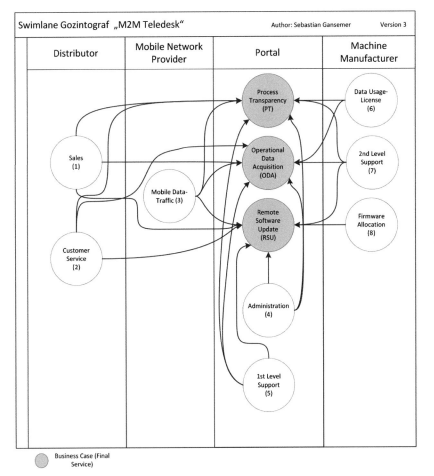

Figure 13.6 Service-links between delivering and receiving services.

Variable costs for the business cases can be calculated using Equation system (13.1).

$$c_1 = a_{11} \cdot b_1 + a_{12} \cdot b_2 + \ldots + a_{1n} \cdot b_n$$
$$\vdots \qquad\qquad\qquad\qquad\qquad (13.1)$$
$$c_m = a_{m1} \cdot b_1 + a_{m2} \cdot b_2 + a_{mn} \cdot b_n,$$

where a_{ij} – amount of service j delivered for service i; b_j – transfer prices of service j; c_i – variable costs of finally receiving service i; m – number of finally receiving services; n – number of delivering services.

The system of linear equations yields relation matrix A=(a_{ij}) and transfer price vector B=(b_j). The vector C=(c_i) of variable costs can be represented by Equation (13.2).

$$C = A \cdot B. \tag{13.2}$$

Using the vector D=(d_i) consisting of sales prices of finally receiving services Equation (13.3) leads to the vector M=(m_i) of gross margin per unit of all business cases, i.e. finally receiving services.

$$M = D - A \cdot B. \tag{13.3}$$

Figure 13.7 exemplifies the input matrix A and vectors B and D with estimated quantities. In matrix A, the rows indicate the business cases PT (row 1), ODA (row 2) and RSU (row 3). Columns represent delivering services indicated as white circles in Figure 13.6. The elements of vector B represent transfer prices of delivering services. Elements of the vector D represent sales prices of the three business cases.

The results of economic analysis are shown in Figure 13.8. Elements of the calculated vector C indicate variable costs of the three business cases. It

$$A = \begin{pmatrix} 1 & 1 & 0.05 & 1 & 1 & 2 & 0 & 0 \\ 1 & 1 & 2.5 & 1 & 2 & 1 & 0.5 & 0 \\ 0 & 1 & 0.1 & 1 & 0.5 & 0 & 0 & 1 \end{pmatrix} B = \begin{pmatrix} 50 \\ 25 \\ 240 \\ 12 \\ 11 \\ 150 \\ 150 \\ 0 \end{pmatrix} D = \begin{pmatrix} 1000 \\ 1000 \\ 100 \end{pmatrix}$$

Figure 13.7 Relation matrix A, transfer price vector B and sales price vector D.

$$C = \begin{pmatrix} 410 \\ 934 \\ 66.5 \end{pmatrix} M = \begin{pmatrix} 590 \\ 66 \\ 33.5 \end{pmatrix}$$

Figure 13.8 Vector of variable costs C and vector of marginal return per unit M.

can be seen that the marginal return per unit is positive for all three business cases with the highest marginal return for business case Process Transparency.

13.7 Communication Security

Securing the communication channel against unauthorized access and manipulation is another important factor which has to be taken into account.

One has to consider the following communication scenarios: communication between machines and portal using mobile networks such as 3G/4G and WLAN, secure remote firmware update and covering dead spots.

The whole security concept is based on asymmetric encryption. Every participant in the communication chain (machines, machine manufacturer, M2M portal, provider) needs a key pair which should be created on the machine to keep the private key on the device.

This security concept was developed in the context of a bachelor thesis at the FH Dortmund [12]. The main target was the use of open and established standards [13, 14].

13.7.1 CA

The central instance of the security structure is a CA (Certificate Authority) which provides services like issuing certificates (by providing a CSR (Certificate Signing Request)), revoking certificates, checking certificates if they are rejected (through CRLs/OCSP). During the key creation process, a CSR is being created which will be passed to the PKI. The CSR is signed and the certificate is sent back to the device (machine).

13.7.2 Communicating On-the-Go

The communication between the machines and the portal is secured by means of a mutually authenticated HTTPS connection. The portal identifies itself to the machine by presenting its certificate and vice versa. During the initiation of the connection, every device has to check the presented certificate by the other part: 1) is the certificate signed by the M2M CA (this prevents man-in-the-middle-attacks)? If yes: 2) check the certificate of the counterpart against the CA if the certificate is revoked or not. This is done by using OCSP or CRLs (as a fallback in case OCSP is failing).

After the connection has been initiated, both partners can communicate securely, while the security of the underlying network(s) (like mobile 2G/3G, WLAN etc.) is no more important.

13.7.3 Covering Dead Spots

In case that a mobile communication is not possible due to lacking availability, the collected data has to be transferred using other methods. Here, other vehicles (such as transportation vehicles) have to deliver the data from the machine within the dead spot to areas with mobile network coverage from where they are sent to the portal. During the transportation, the data has to be secured against manipulation and unauthorized access.

Preparing a data packet for delivery involves the following steps: At first the data is encrypted using the portal's public key. In order to check if the public key is still valid, it is checked with the corresponding certificate against the CA (through OCSP/CRL). This prevents unauthorized access. In the next step, the signature of the encrypted data is created. Therefore, the checksum of the data is calculated and encrypted with the private key of the originating machine. Both the signature (encrypted checksum) and the encrypted data are sent to the vehicle.

The portal checks the signature by decrypting the checksum using the originating machine's public key (key/certificate is checked through OCSP/CRL) and by creating the checksum itself of the data package. If both checksums match, the data has not been manipulated and can be decrypted using the private key of the portal.

13.7.4 Securing WLAN Infrastructures

In the vicinity of a farm, a wireless LAN connection will be used instead of mobile network connection. The M2M project elaborated a reference WLAN network which can be installed on the farm premises. This network is designed and optimized for the M2M system. In order to guarantee that only authorized machines have access to the network, the authentication scheme is based on IEEE 802.1X with a RADIUS using AES/CCMP encryption (IEEE 802.11i RSN). Furthermore, a DHCP/DNS service is provided by the gateway which interconnects the network to the internet and acts as relay to the M2M portal. A machine connects to the M2M wireless network by using its X.509v2 certificate. The certificate is presented to the RADIUS server which performs checks (OCSP/CRL) against the CA whether it is revoked or not and whether the certificate is signed by the M2M CA. The machine itself has to check the RADIUS certificate whether it belongs to the M2M CA in order to avoid rogue access points. If all checks are passed successfully, the RADIUS server grants access to the network.

13.7.5 Firmware Update

It is necessary to periodically apply updates for the software systems on the machines. The updates are passed from the manufacturer of the machine through the portal to the destination machine.

Since the update packages may contain critical and confidential data, the provision of the update package has to be secured accordingly. Because of the file size (100MB and up), asymmetric encryption is not appropriate. Instead, a 256-bit symmetric AES key is generated and is used to encrypt the update. This key is secured using the public key encryption (after checking the corresponding certificate through OCSP/CRL). Here, the public key of the destination machine is used. In the next step, the signature of the encrypted update file is calculated by generating the hash value which then is encrypted with the private key of the manufacturer. Now the signature, the encrypted file and the encrypted AES key are sent to the destination machine.

On the latter, the signature is checked by generating the checksum of the encrypted file and by comparing it with the decrypted checksum. The checksum is decrypted with the public key of the manufacturer. The corresponding certificate has to be checked, too) checksum. If both checksums match, the update did not lose integrity. Finally, the AES key can be decrypt using the private key of the destination machine and the update can be decrypted (Figure 13.9).

13.8 Resume

This paper presents a concept for the optimization of process information chain to improve efficiency in agricultural harvesting process. Machine-to-machine communication plays a central role to synchronize data between diverse process partners.

The information gathered by sensors at agricultural machines plays the central role to build new business models. Business model analysis shows that all parties along the value chain gain good business potential. It has been shown that the three described business models can be operated with positive marginal return per unit under the assumptions made in the project.

However, security issues and business models play an important role for a successful system operation. With the described security measures, system operation can be done ensuring confidentiality, integrity as well as availability.

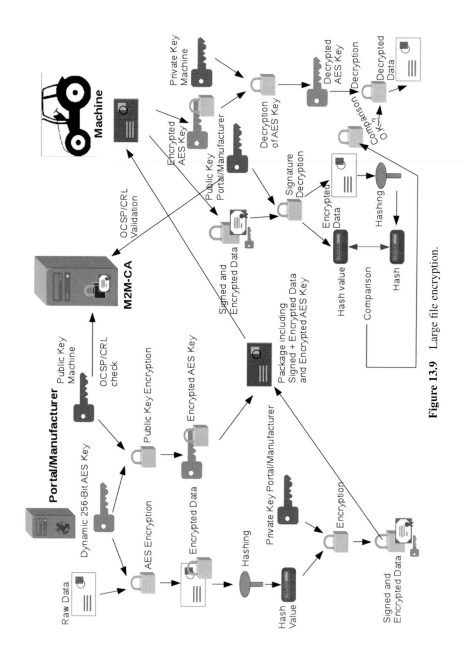

Figure 13.9 Large file encryption.

As the system is designed with an open and generic approach, adoption to other branches such as construction and the opportunity to bring in new functions and business models via third-party software brings additional market potential.

The concepts presented in this paper were developed within the project M2M-Teledesk. The project aims to implement a prototypical system following the concept described above.

13.9 Acknowledgement

The presented work was done in the research project "M2M-Teledesk" funded by the state government of North-Rhine-Westfalia and European Union Fund for regional development (EUROPÄISCHE UNION - Europäischer Fonds für regionale Entwicklung - Investition in unsere Zukunft). Project partners of M2M-Teledesk are University of Applied Sciences and Arts, Dortmund, VIVAI Software AG, Dortmund and Claas Selbstfahrende Erntemaschinen GmbH, Harsewinkel.

References

[1] K. Moummadi, R. Abidar and H. Medromi, 'Generic model based on constraint programming and multi-agent system for M2M services and agricultural decision support', In Multimedia Computing and Systems (ICMCS), 2011:1–6.

[2] G. Wu, S. Talwar, K. Johnsson, N. Himayat and K. Johnson, 'M2M: From Mobile to Embedded Internet', In IEEE Communications Magazine, 2011:36–42.

[3] Y. Daesub, C. Jongwoo, K. Hyunsuk and K. Juwan, 'Future Automotive Insurance System based on Telematics Technology', In 10th International Conference on Advanced Communication Technology (ICACT), 2008:679–681.

[4] A. Juliandri, M. Musida and Supriyadi, 'Positioning cloud computing in machine to machine business models', In Cloud Computing and Social Networking (ICCCSN), 2012:1–4.

[5] V. Goncalves and P. Dobbelaere, 'Business Scenarios for Machine-to-Machine Mobile Applications', In Mobile Business and 2010 Ninth Global Mobility Roundtable (ICMB-GMR), 2010.

[6] Y. Chang, T. Chi, W. Wang and S. Kuo, 'Dynamic software update model for remote entity management of machine-to-machine service capability', In IET Communications Journal, 2012.

[7] S. Blank, G. Kormann and K. Berns, 'A Modular Sensor Fusion Approach for Agricultural Machines', In XXXVI CIOSTA\& CIGR Section V Conference, Vienna, 2011.

[8] M. Mau, 'Supply Chain Management in Agriculture - Including Economics Aspects like Responsibility and Transparency', In X EAAE Congress Exploring Diversity in European Agriculture, 2002.

[9] S. Gansemer, U. Grossmann, B. Horster, T. Horster-Moeller and C. Rusch, 'Machine-to-machine communication for optimization of information chain in agricultural business', In 7th IEEE International Conference on Intelligent Data Acquisition and Advances Computing Systems, Berlin, 2013.

[10] K. Johansson, 'Cost efficient provisioning of wireless access: Infrastructure cost modeling and multi-operator ressource sharing', Thesis KTH School of Electrical Engineering, Stockholm, 2005.

[11] A. Vazsonyi, 'The use of mathematics in production and investory control', Management Science, 1 (1), Jan. 1955.

[12] R.K. Ko, S. S. G. Lee and E. W. Lee, 'Business process management (BPM) standards: a survey', Business Process Mansgement Journal, 15(5):744–791, 2009.

[13] J. Sell, 'Konzeption einer Ende-zu-Ende Absicherung für eine M2M-Telematik Anwendung für die Firma Claas', Thesis FH Dortmund, Dortmund, 2013.

[14] E. Eren and K. Detken, 'Mobile Security. Risiken mobiler Kommunikation und Lösungen zur mobilen Sicherheit', Wien, Carl Hanser Verlag, 2006.

[15] E. Eren and G. Aljabari, 'Virtualization of Wireless LAN Infrastructures', In 6th IEEE Workshop on Intelligent Data Acquisition and Advanced Computing Systems, Prague, 2011.

Index

313

Editor's Biographies

Richard J. Duro received the B.Sc., M.Sc., and Ph.D. degrees in physics from the University of Santiago de Compostela, Spain, in 1988, 1989, and 1992, respectively.

He is currently a Full Professor in the Department of Computer Science and head of the Integrated Group for Engineering Research at the University of A Coruña, Coruña, Spain. His research interests include higher order neural network structures, signal processing, and autonomous and evolutionary robotics.

Yuriy P. Kondratenko, Doctor of Science, Professor, Honour Inventor of Ukraine (2008), Corr. Academician of Royal Academy of Doctors (Barcelona, Spain), Professor of Intelligent Information Systems at Petro Mohyla Black Sea State University, Ukraine. He has received a Ph.D. (1983) and a Dr.Sc. (1994) in Elements and Devices of Computer and Control Systems from Odessa National Polytechnic University. He received several international grants and scholarships for conducting research at Institute of Automation of Chongqing University, P.R.China (1988–1989), Ruhr-University Bochum, Germany (2000, 2010), Nazareth College and Cleveland State University, USA (2003). His research interests include robotics, automation, sensors and control systems, intelligent decision support systems, fuzzy logic, soft computing, elements and devices of computing systems. He is the principal researcher of several international research projects with Spain, P.R. of China et al. and author of more than 120 patents and 12 books (including author's chapters in monographs) published by Springer, World Scientific, Pergamon Press, Academic Verlag etc. He is a member of the GAMM, DAAAM, AMSE UAPL and PBD-Honor Society of International Scholars and visiting lecture at the universities in Rochester, Cleveland, Kassel, Vladivostok and Warsaw.

Author's Biographies

Francisco Bellas received the B.Sc. and M.Sc. degrees in physics from the University of Santiago de Compostela, Spain, in 1999 and 2001, respectively, and the Ph.D. degree in computer science from the University of A Coruña, Coruña, Spain, in 2003.

He is currently a Profesor Contratado Doctor at the University of A Coruña. He is a member of the Integrated Group for Engineering Research at the University of A Coruña. His current research interests are related to evolutionary algorithms applied to artificial neural networks, multiagent systems, and robotics.

Mitru Corneliu Caraivan has received his B.Sc. degree in Automatic Control and Computers in 2009, from Politehnica University of Bucharest, Romania. Following the master thesis in Fakultät für Inginieurwissenshaften, University of Duisburg-Essen, Germany, he earned the Ph.D. degree in Systems Engineering in 2013 from Politehnica University of Bucharest. Since 2009 he is part-time Assistant Professor in the Faculty of Applied Sciences and Engineering, Ovidius University of Constanta, Romania. His main research interests focus on offshore oil and gas automatic control systems on drilling and exploration rigs, programmable logic controller networks, instrumentation data sensors and actuators, systems redundancy and reliability. Since 2010, he has gained experience on offshore jack-up rigs as IT/Electronics Engineer focusing on specific industry equipment: NOV Amphion Systems, Cameron (former TTS-Sense) X-COM Cyber-Chairs 5th gen, drilling equipment instrumentation, PLCs, SBCs, fire and gas alarm systems, satellite communications and networking solutions.

Valentin Dache has received his B.Sc. degree in Automatic Control and Computers in 2008, from the Politehnica University of Bucharest, Romania. Following the master thesis in the same university, he is currently a Ph.D. student in Systems Engineering, having an on-going research which focuses on intelligent buildings management system. Other interests include the field of remotely controlled radiation scanners and non-intrusive inspection systems for customs control and borders security - ROBOSCAN and ROBOSCAN 2M AERIA systems - two times winner of the Grand Prix of Salon International des Inventions de Genève, in 2009 and 2013. Working in the team which won the International Exhibition of Inventions he also owns one invention patent.

Álvaro Deibe Díaz received the M.S. degree in Industrial Engineering in 1994 from the University of Vigo, Spain, and a Ph.D. in Industrial Engineering in 2010 in the University of A Coruña. He is currently Titular de Universidad in the Department of Mathematical and Representation Methods at the same University. He is a member of the Integrated Group for Engineering Research at the University of A Coruña. His research interests include Automation and Embedded Systems.

Alexander A. Dyda was graduated from Electrical Engineering Department of Far-Eastern Polytechnic Institute (Vladivostok) in 1977. He received degree of Candidate of Science (PhD) from Leningrad Electrical Engineering Institute in 1986. From 1979 to 1999 he was Assistant Professor, principal lecturer and Associate Professor of Department of Technical Cybernetics & Informatics. In 1990–1991 and 1995 he was researcher in the Department of System Sciences at "La Sapienza" University (Rome, Italy) and the Department of System Science & Mathematics of Washington University (Saint-Louis, USA), respectively. In 1998 he received degree of Doctor of Technical Sciences (DSc) from Institute of Automation & Control Processes, Russian Academy of Sciences. From 1999 to 2003 he was Chairman of Department of Information Systems. Since 2003 he is Full Professor of Department of Automatic & Information Systems and head of the laboratory of Nonlinear & Intelligent Control Systems at Maritime State University.

Uladzimir Dziomin is a PhD student at Intelligent Information Technology Department of Brest State Technical University. He received MSc and Diploma in Information Technology at BrSTU. The area of his research is Intelligent Control, Autonomous Learning Robots and Distributed Systems.

Chernakova Svetlana Eduardovna was born on August 3, 1967 in the town of Kalinin in the Kalinin Region. She graduated from the Leningrad Polytechnic Institute in 1990, has written more than 80 scientific papers, and possesses considerable knowledge related to development and creation of television systems for different purposes. Area of interest: pattern identification, intelligent systems "human-to-machine", intelligent technology of training by demonstration, mixed (virtual) reality. Place of work: St. Petersburg Institute of Informatics of the Russian Academy of Sciences (SPIIRAS), St. Petersburg E-mail: S_chernakova@rambler.ru Address: 199226, Saint-Petersburg, Nalichnaya Street, H# 36, Building 6, Apt 160

Andrés Faíña received the M.Sc. and Ph.D. degrees in industrial engineering from the University of A Coruña, Spain, in 2006 and 2011, respectively.

He is currently a Postdoctoral Researcher at the IT University of Copenhagen, Denmark. His interests include modular and self-reconfigurable robotics, evolutionary robotics, mobile robotics, and electronic and mechanical design.

Vladimir Golovko, Professor, is a Head of Intelligent Information Technologies Department and Laboratory of Artificial Neural Networks since 2003. He received his PhD degree from the Belarus State University and Doctor of Sciences degree in Computer Science from United Institutes of Informatics Problem of National Academy of Sciences (Belarus). His research interests include Artificial Intelligence, Neural Networks, Autonomous Learning Robots, Intelligent signal processing.

Marcos Miguez González received his master degree in Naval Architecture and Marine Engineering in 2004, from the University of A Coruña, Spain, and his PhD related with the analysis of parametric roll resonance for the same university in 2012. He worked for three years in shipbuilding industry. Since 2007, he is a researcher in the GII (Integrated Group for Engineering Research) of the University of A Coruña, formerly under a scholarship from the Spanish Ministry of Education and nowadays as an assistant professor. His research topics are ship behaviour modelling, parametric rolling and stability guidance systems.

Boris Gordeev received the PhD degree from Leningrad Institute of High-precision Mechanics and Optics, Leningrad, USSR, in 1987 and Doctor of Science degree from Institute of Electrodynamics of National Academy of Sciences of Ukraine in Elements and Devices of Computer and Control Systems in 2011.

He is currently a Professor of Marine Instrumentation Department at National University of Shipbuilding, Ukraine. His current research interest is related to polymetric signals generation and processing for measuring systems, including control systems for continuous measurement of the quantitative and qualitative parameters of the fuels and liquefied petroleum gas. He is the author of more than 150 scientific publications in the above mentioned field.

Anton Kabysh, assistant of Intelligent Information Technologies Department of Brest State Technical University. He is working on Ph.D thesis in Multi-Agent Control and Reinforcement Learning at Robotics Laboratory in BrSTU. Research interest also includes distributed systems, swarm intelligence and optimal control.

Volodymyr Y. Kondratenko received a PhD degree in Applied Mathematics from Colorado University Denver in 2015. He has received a B.S. summa cum laude in Computer Science and Mathematics at Taras Shevchenko Kyiv National University (2008), and a M.S. at University of Colorado (2010) as a Winner of Fulbright Scholarship. He is the author of numerous publications in such fields as Wildfire Modelling and Fuzzy logic, Medallist of the Small Academy of Science of Ukraine (2004, 2nd place) and winner of All-Ukrainian Olympiad in mathematics (2004, 3rd place). His research interests include Computational Mathematics, Probability Theory, Data Assimilation, Computer Modelling, and Fuzzy Logic.

Sauro Longhi received the Doctor degree in Electronic Engineering in 1979 from the University of Ancona, Italy, and post-graduate Diploma in Automatic Control in 1984 from the University of Rome "La Sapienza", Italy. From 1980 to 1981 he held a fellowship at the University of Ancona. From 1981 to 1983 he was with the R&D Laboratory of the Telettra S.p.A., Chieti, Italy, mainly involved in activities of research and electronic design of modulation and demodulation systems for spread spectrum numerical transmission systems. Since 1983 he has been at the Dipartimento di Elettronica e Automatica of the University of Ancona, now Information Engineering Department, of the Università Politecnica delle State Marche-Ancona. From July 2011 to October 2013 he was the Director of this Department. Since November 1, 2013 he is the Rector of the Università Politecnica delle Marche.

Andrei Maciuca obtained his PhD in 2014 from the Department of Automatic Control and Industrial Informatics, University "Politehnica" of Bucharest with a thesis on "Sensor Networks for Home Monitoring of Elderly and Chronic Patients".

Kurosh Madani received his PhD degree in Electrical Engineering and Computer Sciences from University Paris-Sud, Orsay (France) in 1990. In 1995, he received the Doctor Habilitate degree (Dr Hab.) from UPEC. Since 1998 he has worked as Chair Professor in Electrical Engineering at University Paris-Est Creteil (UPEC). Co-initiator of the Images, Signals and Intelligent Systems Laboratory (LISSI/EA 3956), he is head of one of the four research groups of LISSI. His current research interests include bio-inspired perception, artificial awareness, cognitive robotics, human-like robots and intelligent machines. Since 1996 he is elected Academician of the International Informatization Academy. In 1997 he was elected as Academician of the International Academy of Technological Cybernetics.

Kulakov Felix Mikhailovich was born on July 4, 1931 in the town of Nadezhdinsk in Ural Region. He graduated from the Leningrad Polytechnic Institute in 1955, and has written more than 230 scientific papers. He works in the sphere of robot supervision and research of automatic performance of the mechatronic systems. He is an Honored Worker of Science of the Russian Federation, Professor, Chief Research scientist, St. Petersburg Institute of Informatics of the Russian Academy of Sciences (SPII RAS), head of the branch of the department of the "Mechanics of the controlled movement" of the St. Petersburg State University at SPIIRAS. E-mail: kufelix@yandex.ru Address: 198064, Saint-Petersburg, Tikhoretsky Avenue, H# 11, building 4, Apt 50

Andrea Monteriù received the Laurea Degree (joint BSc/MSc equivalent) summa cum laude in Electronic Engineering and the Ph.D. degree in Artificial Intelligence Systems from the Università Politecnica delle Marche, Ancona, Italy, in 2003 and 2006, respectively. His MSc thesis has been developed to the Automation Department of the Technical University of Denmark, Lyngby, Denmark. In 2005 he was a visiting researcher at the Center for Robot Assisted Search & Rescue of the University of South Florida, Tampa, Florida. Since 2005, he is Teaching Assistant of Automatic Control, Automation Systems, Industrial Automation, Modelling and Identification of Dynamic Processes. Since 2007, he has a PostDoc and Research Fellowship at the Dipartimento di Ingegneria dell'Informazione of the Università Politecnica delle Marche, where currently he is a Contract Professor.

Alexander Nakonechniy received the B.Sc. and M.Sc. degrees in High Voltages Electro-physics from National University of Shipbuilding, Ukraine, in 2007 and 2009 respectively.

He is currently a post-graduate student at Marine Instrumentation Department of National University of Shipbuilding, Ukraine. His current research interests are polymetric measurements, sensors and sensors networks, generation of short pulses. He is the author of several scientific publications in the above fields.

Sergey I. Osadchy Doctor of Science, Professor, Chief of Production Process Automatic Performance Department at Kirovograd National Technical University, Ukraine. He has received a Ph.D. (1987) and a Dr.Sc. (2013) in Automatic Performance of Control Processesfrom National Technic University of Ukraine "Kiev Polytechnic Institute". His research interests include robotics, automation, sensors and optimal control systems synthesis, analysis and identification, underwater supercavitating vehicle stabilization and control. He is the author of more than 140 scientific articles, 25 patents and 4 books.

Dmitry A. Oskin was graduated from Electrical Engineering Department of Far-Eastern State Technical University (Vladivostok) in 1997. He received degree of Candidate of Science (PhD) in 2004. From 2000 to 2005 he was an Assistant Professor of Department of Information Systems. Since 2005, he is Associate Professor of Department of Automatic & Information Systems and Senior researcher of the laboratory of Nonlinear & Intelligent Control Systems at Maritime State University.

Fernando López Peña received the Master degree in Aeronautical Engineering from the Polytechnic University of Madrid, Spain, in 1981, a Research Master from the von Karman Institute of Fluid Dynamics, Belgium, in 1987, and the Ph.D. degree from the University of Louvain, Belgium, in 1992. He is currently a professor at the University of A Coruña, Spain. He authored about 100 papers on peer reviewed journals and international conferences, holds five patents, and has leaded more than 80 research projects. His current research activities are related to intelligent hydrodynamics and aerodynamics design and optimization, flow measurement and diagnosis, and signal and image processing.

Dan Popescu obtained the MSc degree in Automatic Control in 1974, MS in mathematics in 1980, PhD degree in Electrical Engineering in 1987. He is currently Full Professor, in the Department of Automation and Industrial Informatics, University "Politehnica" of Bucharest, head of the laboratory Artificial Vision and responsible for master program Complex Signal

Processing in Multimedia Applications. He is PhD adviser in the field of Automatic Control. Current scientific areas are: equipments for complex measurements, data acquisition and remote control, wireless sensor networks, alerting systems, pattern recognition and complex image processing, interdisciplinary approaches. He is author of 15 books, more than 120 papers, and director of five national research grants. He is IEEE member, SRAIT member, and he received the 2008 IBM Faculty Award. His research interests lay in the field of wireless sensor networks, image processing and reconfigurable computing.

Dominik Maximilián Ramík received his PhD in 2012 in signal and image processing from University of Paris-Est, France. Member of LISSI laboratory of University Paris-Est Creteil (UPEC), his current research topic concerns processing of complex images using bio-inspired artificial intelligence approaches and consequent extraction of semantic information with use in mobile robotics control and industrial processes supervision.

Christian Rusch studied Mechatronics between 1999 and 2005 and graduated from the Technical University of Braunschweig. Then he started his scientific career as Research Scientist at the Technical University of Berlin. The Title of his PhD Thesis is "Analysis of data security of self-configuring radio networks on mobile working machines illustrated by the process documentation". He finished the PhD (Dr. –Ing.) in 2012. In 2011 he moved in to the industry and worked for CLAAS as project manager with focus on wireless communication for mobile working machines. Since 2014 he is system engineer in CLAAS Electronic Systems. Furthermore, he is project leader of the standardization group "Wireless ISOBUS Communication".

Félix Orjales Saavedra received the master degree in Industrial Engineering in 2010 from the University of León, Spain. He is currently working on his Ph.D. degree in the Department of Industrial Engineering at the University of A Coruña, Spain, related with collaborative unmanned aerial vehicles. He is a researcher at the Integrated Group for Engineering Research, and his main research topics include autonomous systems, robotics, and electronic design.

Christophe Sabourin received his PhD in Robotics and Control from University of Orleans (France) in November 2004. Since 2005, he has been a researcher and a staff member of Images, Signals and Intelligent Systems Laboratory (LISSI/EA 3956) of University Paris-Est Creteil (UPEC).

His current interests relate to areas of complex and bio-inspired intelligent artificial systems, cognitive robotics, humanoid robots, collective and social robotics.

Valentin Sgârciu is full professor in the Faculty of Automatic Control and Computers, Politehnica University of Bucharest, Romania since 1995 and PhD coordinator in the same university since 2008. His teaching activity includes undergraduate and postgraduate courses: "Sensors and Transducers", "Data Processing", "Transducers and Measuring systems", "Reliability and Diagnosis", "Security of Informatics Systems" and "Product and System Diagnosis". His research activity include the fields of intelligent buildings, intelligent measurements, wireless sensor networks, software engineering, operating systems, control system theory, e-learning, distributed heterogeneous systems and middleware technologies. He has over 100 scientific papers presented in national and international symposiums and congresses, published in the associated preprints or in specialized Romanian and international journals. He is first author of 16 books and owns one invention patent.

Daniel Souto received the M.Sc. degree in industrial engineering from the University of A Coruña, Coruña, Spain, in 2007. He is working towards the Ph.D. degree in the Department of Industrial Engineering at the same university.

He is currently a Researcher at the Integrated Group for Engineering Research. His research activities are related to automatic design and mechanical design of robots.

Grigore Stamatescu is currently Assistant Professor within the Department of Automatic Control and Industrial Informatics, University "Politehnica" of Bucharest – Intelligent Measurement Technologies and Transducers laboratory. He obtained his PhD in 2012 from the same institution with a thesis on "Improving Life and Work with Reliable Wireless Sensor Networks". He is a member of the IEEE Instrumentation and Measurement and Industrial Electronics societies and has published over 50 papers in international journals and indexed conference proceedings. His current research interests lay in the fields of intelligent sensors, embedded networked sensing and information processing.

Ralf Stetter is a Professor in the Department of Mechanical Engineering at the Hochschule Ravensburg-Weingarten since 2004. His Teaching Area is "Design and Development in Automotive Technology". He received his PhD and Diploma in Mechanical Engineering from the Technische Universität München (TUM). He is currently Vice-Dean of the Department of Mechanical Engineering. He works as Project Manager in the Steinbeis-Transfer-Centre "Automotive Systems". He was Team Coordinator at Audi AG, Ingolstadt in the Automotive Interior Product Development.

Mircea Strutu obtained his PhD in 2014 from the Department of Automatic Control and Industrial Informatics, University "Politehnica" of Bucharest with a thesis on "Wireless Networks for Environmental Monitoring and Alerting based on Mobile Agents".

Anna S. Timoshenko, Lecturer in Department of Information Technology at Kirovograd Flight Academy of National Aviation University. She graduated from the State Flight Academy with honors in 2002. She graduated from the Kirovograd National Technical University in 2006. Then she took post-graduate studies in Kirovograd Flight Academy of National Aviation University (2009–2012). She is the author of numerous publications in such fields as Navigation and Motion Control, Automation and Robotics.

Blanca María Priego Torres received the title of Telecommunications Engineer in 2009 from the University of Granada, Spain. In 2011, she obtained the Master's Degree in Information and Communications Technologies in Mobile Networks from the University of A Coruña, Spain. She is currently working in her PhD as a member of the Integrated Group for Engineering Research at the University of A Coruña. Her research interests include the analysis, processing and interpretation of signals and multi-dimensional images in relation to the Industrial and Naval Field, and through the application of techniques based on new neural-network structures, soft computing and machine learning.

Yuriy Zhukov received the PhD and Doctor of Science degrees from Nikolayev Shipbuilding Institute (now NUOS), Ukraine, in 1981 and 1994, respectively.

He is currently a Professor, Chief of Marine Instrumentation Department and Neoteric Naval Engineering Institute at National University of Shipbuilding, Ukraine. He is the author of more than 250 scientific publications in fields of ship design and its operational safety, dynamic systems behavior monitoring and control, decision making support systems and artificial intelligence, etc. His current research interest is related to intelligent multi-agent sensory systems application in the above mentioned fields.

Alexey Zivenko received the PhD degree in Computer Systems and Components from Petro Mohyla Black Sea State University, Ukraine, in 2013.

He is currently an Associate Professor of Marine Instrumentation Department at National University of Shipbuilding, Ukraine. His current research interests are polymetric measurements, non-destructive assay of liquids, intellectual measurement systems and robotics, data acquisition systems and data mining. He is the author of more than 20 scientific publications in the above mentioned fields.

Valerii A. Zozulya, Associate Professor in Department of automation of production processes at Kirovograd National Technical University, Ukraine. He has received a Ph.D. (1999) and received the academic rank of associate professor (2003) at Kirovograd National Technical University. He is the author of numerous publications in such fields as motion control, robotics and optimal control systems synthesis, analysis and identification.